AF576141

Recent Methods from Statistics and Machine Learning for Credit Scoring

Anne Kraus

München 2014

Bibliografische Information der Deutschen Nationalbibliothek
Die Deutsche Nationalbibliothek verzeichnet diese Publikation in der Deutschen Nationalbibliografie; detaillierte bibliografische Daten sind im Internet über http://dnb.d-nb.de abrufbar.
1. Aufl. - Göttingen: Cuvillier, 2014
Zugl.: München, Univ., Diss., 2014

Nonnenstieg 8, 37075 Göttingen
Telefon: 0551-54724-0
Telefax: 0551-54724-21
www.cuvillier.de

1. Auflage, 2014
Gedruckt auf umweltfreundlichem, säurefreiem Papier aus nachhaltiger Forstwirtschaft

ISBN 978-3-95404-736-9
eISBN 978-3-7369-4736-8

Recent Methods from Statistics and Machine Learning for Credit Scoring

Anne Kraus

Dissertation
an der Fakultät für Mathematik, Informatik und Statistik
der Ludwig–Maximilians–Universität
München

vorgelegt von
Anne Kraus
aus Schweinfurt

München, den 10. März 2014

Erstgutachter: Prof. Dr. Helmut Küchenhoff
Zweitgutachter: Prof. Dr. Martin Missong
Tag der Disputation: 22. Mai 2014

Mein besonderer Dank gilt

- Prof. Dr. Helmut Küchenhoff, für die Möglichkeit bei ihm zu promovieren, die Begeisterung für das Thema und die hervorragende Betreuung in den letzten Jahren
- Prof. Stefan Mittnik, Ph.D., für die Zweitbetreuung im Rahmen des Promotionsprogramms
- Prof. Dr. Martin Missong, für die Bereitschaft, das Zweitgutachten zu übernehmen
- allen Doktoranden und Mitarbeitern am Institut für Statistik, für die freundliche Aufnahme und Unterstützung, ganz besonders Monia Mahling
- meinem Arbeitgeber, für die Möglichkeit der Freistellung
- meinen Kollegen, für ihr Interesse und viele gemeinsame Mittagspausen
- meinen Freunden, für viel Verständnis und Motivation, allen voran Karin Schröter
- meinen Geschwistern Eva Schmitt und Wolfgang Kraus samt Familien, für Aufmunterung und Ablenkung
- meinen Eltern, für ihre grenzenlose Unterstützung in jeglicher Hinsicht
- Martin Tusch, für unendlichen Rückhalt

Abstract

Credit scoring models are the basis for financial institutions like retail and consumer credit banks. The purpose of the models is to evaluate the likelihood of credit applicants defaulting in order to decide whether to grant them credit. The area under the receiver operating characteristic (ROC) curve (AUC) is one of the most commonly used measures to evaluate predictive performance in credit scoring. The aim of this thesis is to benchmark different methods for building scoring models in order to maximize the AUC. While this measure is used to evaluate the predictive accuracy of the presented algorithms, the AUC is especially introduced as direct optimization criterion.

The logistic regression model is the most widely used method for creating credit scorecards and classifying applicants into risk classes. Since this development process, based on the logit model, is standard in the retail banking practice, the predictive accuracy of this proceeding is used for benchmark reasons throughout this thesis.

The AUC approach is a main task introduced within this work. Instead of using the maximum likelihood estimation, the AUC is considered as objective function to optimize it directly. The coefficients are estimated by calculating the AUC measure with Wilcoxon–Mann–Whitney and by using the Nelder–Mead algorithm for the optimization. The AUC optimization denotes a distribution-free approach, which is analyzed within a simulation study for investigating the theoretical considerations. It can be shown that the approach still works even if the underlying distribution is not logistic.

In addition to the AUC approach and classical well-known methods like generalized additive models, new methods from statistics and machine learning are evaluated for the credit scoring case. Conditional inference trees, model-based recursive partitioning methods and random forests are presented as recursive partitioning algorithms. Boosting algorithms are also explored by additionally using the AUC as a loss function.

The empirical evaluation is based on data from a German bank. From the application scoring, 26 attributes are included in the analysis. Besides the AUC, different performance measures are used for evaluating the predictive performance of scoring models. While classification trees cannot improve predictive accuracy for the current credit scoring case, the AUC approach and special boosting methods provide outperforming results compared to the robust classical scoring models regarding the predictive performance with the AUC measure.

Zusammenfassung

Scoringmodelle dienen Finanzinstituten als Grundlage dafür, die Ausfallwahrscheinlichkeit von Kreditantragstellern zu berechnen und zu entscheiden ob ein Kredit gewährt wird oder nicht. Das AUC (area under the receiver operating characteristic curve) ist eines der am häufigsten verwendeten Maße, um die Vorhersagekraft im Kreditscoring zu bewerten. Demzufolge besteht das Ziel dieser Arbeit darin, verschiedene Methoden zur Scoremodell-Bildung hinsichtlich eines optimierten AUC Maßes zu „benchmarken". Während das genannte Maß dazu dient die vorgestellten Algorithmen hinsichtlich ihrer Trennschärfe zu bewerten, wird das AUC insbesondere als direktes Optimierungskriterium eingeführt.

Die logistische Regression ist das am häufigsten verwendete Verfahren zur Entwicklung von Scorekarten und die Einteilung der Antragsteller in Risikoklassen. Da der Entwicklungsprozess mittels logistischer Regression im Retail-Bankenbereich stark etabliert ist, wird die Trennschärfe dieses Verfahrens in der vorliegenden Arbeit als Benchmark verwendet.

Der AUC Ansatz wird als entscheidender Teil dieser Arbeit vorgestellt. Anstatt die Maximum Likelihood Schätzung zu verwenden, wird das AUC als direkte Zielfunktion zur Optimierung verwendet. Die Koeffizienten werden geschätzt, indem für die Berechnung des AUC die Wilcoxon Statistik und für die Optimierung der Nelder-Mead Algorithmus verwendet wird. Die AUC Optimierung stellt einen verteilungsfreien Ansatz dar, der im Rahmen einer Simulationsstudie untersucht wird, um die theoretischen Überlegungen zu analysieren. Es kann gezeigt werden, dass der Ansatz auch dann funktioniert, wenn in den Daten kein logistischer Zusammenhang vorliegt.

Zusätzlich zum AUC Ansatz und bekannten Methoden wie Generalisierten Additiven Modellen, werden neue Methoden aus der Statistik und dem Machine Learning für das Kreditscoring evaluiert. Klassifikationsbäume, Modell-basierte Recursive Partitioning Methoden und Random Forests werden als Recursive Paritioning Methoden vorgestellt. Darüberhinaus werden Boosting Algorithmen untersucht, die auch das AUC Maß als Verlustfunktion verwenden.

Die empirische Analyse basiert auf Daten einer deutschen Kreditbank. 26 Variablen werden im Rahmen der Analyse untersucht. Neben dem AUC Maß werden verschieden Performancemaße verwendet, um die Trennschärfe von Scoringmodellen zu bewerten. Während Klassifikationsbäume im vorliegenden Kreditscoring Fall keine Verbesserungen erzielen, weisen der AUC Ansatz und einige Boosting Verfahren gute Ergebnisse im Vergleich zum robusten klassischen Scoringmodell hinsichtlich des AUC Maßes auf.

Contents

List of Figures

List of Tables

Chapter 1

Introduction

1.1 Credit Scoring

Customers apply for consumer credit loans for many reasons, including flat-panel tvs, long-distance trips, luxury cars, relocation and family needs. The amount of consumer credits is lower than for real estate financing and involves several thousand euros, normally less than a hundred thousand. In this kind of credit business, personal information of the applicants are the main basis for the assessment of risk. This is in contrast to real estate financing, where higher credit amounts, lower interest rates and equity capital are the characteristics.

Since customer demand for personal loans has increased in the last decades, the consumer credit market evolved to become an important sector in the financial field and today represents a high-volume business. The UK and Germany are the countries with the largest total amount of consumer credits in Europe (Thomas, 2009). The German retail bank, from which the data is used in this thesis, receives about a million applications per year. These developments in the retail credit market requires automatic, fast and consistent decisions and processes to handle the huge amount of applications. The use of credit scoring models is now a key component in retail banking.

The development of so-called scorecards therefore represents the core competence of a retail bank's risk management when assessing the creditworthiness of an individual. The credit scorecards are embedded in the whole decision process, where other parts like budget account, the examination of the revenue and expenditure accounts and policy rules, are also relevant. These rules lead directly to the rejection of an applicant. Logistic regression is the main method applied in the banking sector to develop the scoring models. The performance of these models is essential, and improvements in the predictive accuracy can lead to significant future savings for the retail bank. The analysis of different models and algorithms to develop scorecards is therefore substantial for credit banks.

Since the market is changing rapidly, new statistical and mathematical methods are required for optimizing the scoring problem to decide on the question of whom to lend credit to. In recent years, many quantitative techniques have been used to examine predictive power in credit scoring. Discriminant analysis, linear regression, logistic regression, neural networks, K-nearest neighbors, support vector machines and classification trees cover the range of different surveys (Thomas et al., 2005). An overview of publications is given in Thomas (2000) and Crook et al. (2007). For instance, Baesens et al. (2003) compare different classification techniques for credit scoring data where neural networks and least-squares support vector machines yield good results, but the classical logistic regression model still performs very well for credit scoring. In the meantime, new developments in statistics and machine learning raise the question of wether or not newer algorithms would perform better in credit scoring than the standard logistic regression model.

The area under the curve (AUC), based on the receiver operating characteristic (ROC) curve, is the most widely applied performance measure in practice for evaluating scoring models. However, classification algorithms are not necessarily optimal with respect to the AUC measure. This is the initial point for the idea to introduce the AUC as direct optimization criterion in this thesis. Since the AUC is a sum of step functions, the Nelder–Mead algorithm (Nelder and Mead, 1965) is used for the optimization, which represents a derivative-free and direct search method for unconstrained optimization of non-smooth functions. Moreover, the Wilcoxon statistic (Wilcoxon, 1945) is used for calculating the AUC measure. This novel approach is presented for the retail credit scoring case, and the properties of the algorithm are analyzed within a simulation study.

Recursive partitioning methods are very popular in many scientific fields, like bioinformatics or genetics, and represent a non-parametric approach for classification and regression problems. Since classification trees show rather poor results for prediction, random forests are especially prominent representatives of recursive partitioning algorithms leading to high predictive accuracy. Random forests belong to the ensemble methods, that overcome the instability of single classifiers by combining a whole set of single trees. Since the main task of this thesis is to improve the predictive accuracy of scoring models, standard recursive partitioning methods, and especially recent methodological improvements of the algorithms, are evaluated for the credit scoring problem. The new extensions overcome the problems of biased variable selection and overfitting in classification trees and random forests. In addition, an interesting approach is to combine classification trees with the classical logit model. This model-based recursive partitioning approach is also evaluated for the retail credit data with respect to the AUC performance.

The idea of boosting methods is to combine many weak classifiers in order to achieve a high classification performance with a strong classifier. Boosting algorithms are representatives of the machine learning area and offer a huge amount of different approaches. Apart from the classical AdaBoost with classification trees as so-called weak learners, it is possible to estimate a logit model by using linear base learners and the negative binomial log-likelihood loss within the component-wise gradient boosting. This boosting framework comes along with variable selection within the algorithm and more interpretability for the results. Since boosting methods are applied to different problems in other scientific fields, the aim here is to analyze these new developments in the retail credit sector for improving scoring models.

For the evaluation of recent methods from statistics and machine learning for the credit scoring case, the investigations in this thesis are based on data from a German bank. I am grateful to this bank for providing this data. Many surveys in this scientific field investigate well-known and often used public data sets for their empirical evaluation. A German credit data set with a 1,000 observations or a Japanese retail credit risk portfolio are prominent examples.[1]

As outlined above, the main focus of this thesis is the improvement of scoring models in the retail banking sector. It is a statistical point of view on credit scoring with concentration on recent methodological developments. The presented credit scoring data is evaluated with many different statistical methods rather than focusing on one specific model. The AUC is highlighted as performance measure and direct objective function. Moreover, different performance measures are presented for the analysis. In addition to the evaluation and the performance comparison of the presented algorithms, another aim is to stress the pros and cons of the proposed methods and to discuss them in the credit scoring context.

1.2 Scope of the Work

The scope of this work is to benchmark recent methods from statistics and machine learning for creating scoring models in order to maximize the AUC measure. To analyze the topic, this thesis is structured as follows:

Chapter 2 starts with the description of the AUC as a measure of performance in credit scoring, gives a short overview of further performance measures and continues with an overview of the credit scoring data used for the evaluation.

Chapter 3 presents a short introduction to the classical logit model, followed by a short summary of the classical scorecard development process in the retail banking practice.

Chapter 4 introduces a novel approach that uses the AUC as direct optimization criterion, and also includes some theoretical considerations and a simulation study for analyzing the properties of the proposed procedure.

Chapter 5 applies the generalized additive model for the credit scoring case as advancement of the classical logit model and representative of the well-known classical methods.

In Chapter 6 and 7, new methods from machine learning are investigated for the credit scoring case. In Chapter 6, classification trees, random forests and model-based recursive partitioning algorithms are presented and evaluated in the credit scoring context.

[1]Both available at `http://archive.ics.uci.edu/ml/` of the University of California-Irvine (UCI) Repository.

In Chapter 7, the evaluation continues with various boosting methods where different loss functions and base learners are investigated. The AUC is especially used as a loss function within the boosting framework and analyzed for the credit scoring case.

Finally, the most important results are highlighted in the Chapter 8 summary. Concluding remarks and recommendations are given in the final chapter by additionally outlining issues for further research.

Computational aspects are given in the Appendix B where statistical software details are presented (B.1), and exemplary R-Codes are provided for the AUC approach and the simulation study in Chapter 4 (B.2). Additional graphics and result tables are included in the Appendix A.

Parts of Chapters 2 to 4 and 7 are published in an article in the Journal of Risk Model Validation (Kraus and Küchenhoff, 2014).

Chapter 2

Measures of Performance and Data Description

2.1 Receiver Operating Characteristic and Area under the Curve

A great variety of performance measures exists in credit scoring. Kullback divergence, Kolmogorov–Smirnov statistic (KS) and information value denote some of the measures used in this area (Anderson, 2007). The H measure (Hand, 2009) or the error rate are other performance measures used to evaluate scoring models. In this thesis, the focus is on the Receiver Operating Characteristic (ROC) and the related area under the curve (AUC) as the most widely applied performance measures in credit scoring practice. By far, these measures, including the Gini coefficient respectively, are the most important criteria in the retail banking sector. Further measures for discrimination, probability prediction, and categorical forecasts are also presented. The AUC is used as a measure of predictive accuracy for evaluating the different methods and algorithms in the credit scoring context. It is especially introduced as direct optimization criterion.

Primarily, ROC graphs have a long history of describing the tradeoff between hit rates and false alarm rates in signal detection theory (Swets, 1996; Swets et al., 2000). Thereafter, ROC graphs have been mainly used in medical decision making. Pepe (2003) considers, for example, the accuracy of a diagnostic test for the binary variable of the disease status (disease and non-disease). In recent years, the ROC analysis and the AUC have been increasingly used in the evaluation of machine learning algorithms (Bradley, 1997). According to Provost and Fawcett (1997), simple classification accuracy is often a poor metric for measuring performance so the ROC analysis gained more impact. For instance, one-dimensional summary measures, like the overall misclassification rate or odds ratio, can lead to misleading results and are rarely used in practice (Pepe et al., 2006). Since the consequences of false-negative and false-positive errors are hard to quantify, it is common to draw both dimensions (tp rate and fp rate) into account (Pepe et al., 2006).

The survey of the credit scoring case denotes a binary problem regarding whether the customers default in a specific time period or the applicants pay regularly (cf. Section 2.2). The following two classes are examined:

- *default*, i.e., a customer fails to pay installments and gets the third past due notice during the period of 18 months after taking out the loan
- *non-default*, i.e., a customer pays regular installments during the period of 18 months after taking out the loan.

Due to this two class problem, two classes are used for the description of the ROC graph. For the multi-class ROC graphs, I reference the explanation of Hand and Till (2001). The following descriptions are based on Pepe (2003) and Fawcett (2006).
For classification models with a discrete outcome for prediction, different cases arise. If a default is correctly classified and predicted as a default, it is a *true positive*; while a non-default wrongly predicted as a default is counted as a *false positive.* Accordingly, the following parameters are computed:

$$tp\ rate = \frac{defaults\ correctly\ classified\ (tp)}{total\ defaults\ (p)} \tag{2.1}$$

$$fp\ rate = \frac{non\ defaults\ incorrectly\ classified\ (fp)}{total\ non\ defaults\ (n)} \tag{2.2}$$

Plotting the fraction of the correctly classified defaults (tp rate - TPR) versus the incorrectly classified non-defaults (fp rate - FPR) gives rise to the ROC graph. Figure 2.1 shows an example of ROC curves for three different default criteria. The diagonal describes a model with no predictive information, while the perfect model would imply a curve directly tending to the point $(0, 1)$. The perfect model would imply that there exists a score where all defaults have scores below this value, and the non-defaults above this value, respectively.
In the analysis, the main interest is on classifiers that do not produce a discrete good or bad decision, but a probability of default. Thresholds are used to create the ROC graph for these scoring classifiers. The threshold denotes a binary classifier for each special score, so for each threshold value one point in the ROC space can be evaluated. An important advantage of ROC curves is that they are insensitive to changes in the proportion of defaults to non-defaults (Fawcett, 2006).

Supplementary to the curves, numerical indices for the ROC graphs are often used. An important measure is the area under the ROC curve (AUC), which ranges between 0 and 1. As the name indicates this denotes the area under the curve presented in Figure 2.1. A perfect model has an AUC value of 1, while an uninformative scoring classifier has a value of 0.5, respectively. Normally scoring systems in practice have a value in-between.

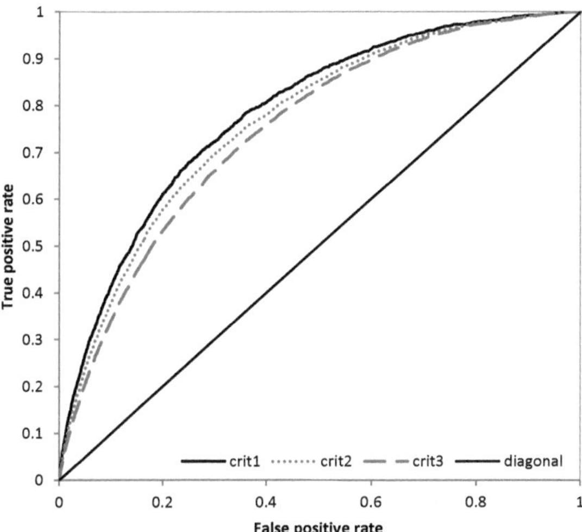

Figure 2.1: Example for ROC curves with three different criterions of default (crit1, crit2, crit3) by plotting the true positive rate versus the false positive rate.

An equivalent transformation to the AUC denotes the Gini coefficient with the definition $Gini = 2 \cdot AUC - 1$.

Since the AUC measure is used as well as direct optimization criterion, it is important to highlight the relationship to the well-studied Wilcoxon statistic (Hanley and McNeil, 1982). Assuming two samples of non-defaults n_{nd} and defaults n_d, all possible comparisons of the scores from each sample correspond to the following rule:

$$S(x_{nd}, x_d) = \begin{cases} 1 & \text{if } x_{nd} > x_d \\ 0.5 & \text{if } x_{nd} = x_d \\ 0 & \text{if } x_{nd} < x_d \end{cases} \tag{2.3}$$

where x_{nd} and x_d are the scores from the non-defaults and defaults, respectively. By averaging over the comparisons, the AUC can be written as follows:

$$AUC = \frac{1}{n_{nd} \cdot n_d} \sum_{1}^{n_{nd}} \sum_{1}^{n_d} S(x_{nd}, x_d) \tag{2.4}$$

This definition of the AUC value will be used for the AUC approach proposed in Chapter 4. Further explanations, especially concerning the confidence intervals, can be found in Hanley and McNeil (1982). Frunza (2013) covers the Gini coefficient in the context of credit risk model validation. For the analysis of areas under the ROC curves, DeLong et al.

(1988) propose a nonparametric approach by using the theory of generalized U-statistics to test whether the difference is significant.

The main objectives for a bank when lending consumer credits are to maximize the profit and reduce losses if a customer can not repay a loan. The cut-off defines the threshold denoting which customers are accepted and rejected, respectively. The ROC curve can also be used to display the influence of changes in the cut-off levels on various business measures of the loan portfolio like profit or loss (Thomas, 2009). But the business perspective goes beyond the scope of this thesis, since the focus lies on the statistical methods and the statistical model estimation with new algorithms. For that reason, I reference Thomas (2009) for further readings in the business aspects of the ROC curve.

In spite of its popularity, critical aspects remain for the AUC measure, such as the fact that it ignores the probability values and only considers the order of the scores (Ferri et al., 2005). Therefore, further performance measures are evaluated for the current credit scoring case.

Hand (2009) discusses some weaknesses of the AUC, such as the incoherency in terms of misclassification costs, and proposes the H measure as an alternative. He derives a linear relationship between AUC and expected minimum loss, where the expectation is taken over a distribution of the misclassification cost parameter that depends on the model under consideration (Flach et al., 2011). The H measure is derived by replacing this distribution with a Beta$(2, 2)$ distribution.

The KS curve charts the empirical cumulative distribution function percentages for defaults and non-defaults against the score. The KS statistic is defined by the maximum absolute difference between the two curves (Anderson, 2007). The above mentioned measures investigate the discrimination of the scoring models.

The minimum error rate (MER) is used to measure categorical forecasts. The performance metrics on comparing the classifier's predicted classes and their true labels are described previously in the context of the ROC-curve. A widely used summary of the so-called misclassification counts is the error rate (ER), which defines the total misclassification count (the sum of the false negatives FN and false positives FP) divided by the number of observations, i.e., $ER = \frac{FN+FP}{n}$. The error rate depends on a special threshold t to produce class labels since most of the classifiers produce a probability prediction. The minimum error rate (MER) corresponds to the value of t that achieves the minimum ER(t) over the test dataset (Anagnostopoulos et al., 2012).

To consider the probability prediction of a scorecard, the Brier score (Brier, 1950) is analyzed. The original definition of Brier (1950) is

$$BS = \frac{1}{n}\sum_{j=1}^{r}\sum_{i=1}^{n}(p_{ij} - Y_{ij})^2 \tag{2.5}$$

where p_{ij} denotes the forecast probabilities, Y_{ij} takes the value 0 or 1 according to whether the event occured in class j or not and r defines the possible classes ($r = 2$ for default and non-default). The Brier score defines, therefore, the expected squared difference of the predicted probability and the response variable. The lower the Brier score of a model, the better is the predictive performance. However, the use of the Brier score is critical since it depends on direct probability predictions.

2.2 Data Description

The data used in this thesis was provided by a German retail bank that specializes in lending consumer credits. The application process of this bank is fully automated, thus offering a very large database with a wide range of information and a long data history. For the analysis, I concentrate on a part of the overall portfolio that guarantees representative results and also protects the bank's business and financial confidentiality.

The focus lies in the application scoring data because the decision of whether to grant credit to an applicant is the key issue in retail banking. If the promise of a loan is given and the money is paid out to a customer, the credit becomes part of the bank portfolio. As a result, the bank cannot cancel the decision. Behavioral scoring is important for forecasting the portfolio risk and is, therefore, relevant for the capital resources demanded by Basel II. Since the information is derived from a real application process, the data quality is very good. Because the customers are interested in taking out a loan, they are willing to divulge their personal information. The verification process inside the bank, where bank employees review the information, guarantees the accuracy of the data. Customers who provided wrong specifications do not receive credit unless the information is corrected. Missing data, therefore, poses a minor problem for the presented credit scoring case.

On the one hand, the most important variables are personal data from the customers, including income, marital status, age and debt position. Information from credit agencies completes the data set while applicants contractually permit the bank to obtain this data.

On the other hand, the payment history of consumers is essential for credit scoring, because this information enables the estimation of credit scoring models. Twenty six attributes are considered regarding process-related limitations for this specific bank. While variables like contract period or the amount of credit are good attributes to help predict a credit default, the decision of whether to grant credit has to be finalized before the customer decides about these variables. I note that these limitations present realistic empirical conditions for this bank. In other business models, for instance, loans are granted after the length of the loan is determined by the customer.

From the 26 attributes, four variables are categorical, while the other 22 covariates are of numerical order. For building credit scoring models, it is quite common to categorize the variables by building classes. Some explanations are given in Section 3.2 by describing the scorecard development process. If in the following variables are denoted as 'classified' or 'categorized', this always refers to the 22 numerical covariates. The remaining four variables are originally categorical and do not need further classification.

For model estimation, there are various options to define the outcome variable. A withdrawal of the credit can be denoted via a default as well as a past due notice. Moreover, it is essential to return to past events for observing the default. This time span can vary

considerably, for example, from 12 to 30 months.
The selected time span is always a tradeoff between actuality and the rate of defaults. For the analysis, the event of sending a third past due notice is assessed as the definition of default and a time span of 18 months is chosen (cf. Section 2.1).

Loans that are past due for more than 90 days can be classified as default as per the Basel II definition (Basel Committee on Banking Supervision, 2004). Since a time span of 12 months is often used for modeling predictive power, the choice of the time span depends on the portfolio structure. For differing customers who fail at the beginning and those failing after several months, survival analysis approaches are used for building scoring models. Banasik et al. (1999) and Stepanova and Thomas (2002) cover these topics in detail.

Due to confidentiality reasons, the descriptive analyses for the variables are not presented. For the evaluation, three different samples for training, test, and validation are used (shown in Table 2.1).

Data	All	Defaults	Defaults in %
trainORG	65818	1697	2.58
test	138189	3634	2.63
validation	72251	1878	2.60

Table 2.1: Data sets for training, test, and validation analyses, where the test and validation data represent out-of-time samples.

The default rates vary around 2.6%, a level that is characteristically low for the credit scoring market. As mentioned above, this quantity is an important reason for choosing a horizon of 18 months and the event of sending the third past due notice. As all three samples are drawn from different time intervals, the investigations on the test and validation samples are out of time.

For test purposes, additionally two samples are generated with randomly created default rates, as shown in Table 2.2. In practice, this procedure is common. The non-defaults are reduced to receive the specified default rate of 20 and 50%. These samples are generated from the original training sample (trainORG) of 65,818 applications.

Training data sets	All	Defaults in %
train020	8485	20
train050	3394	50

Table 2.2: Training data sets with simulated default rates of 20% and 50% drawn from the original training sample (trainORG) by randomly reducing the non-defaults.

The new methods of machine learning presented in this thesis are trained on a training sample, while the different tuning parameters are tuned according to the outcomes on the test sample. The validation sample is finally used for validating the results.

Chapter 3

Logistic Regression

3.1 A Short Introduction to Logistic Regression

In retail banking, logistic regression is the most widely used method for classifying applicants into risk classes. Good interpretability of the results and simple explanation are the most important merits of this procedure (Hand and Kelly, 2002). Another advantage results from directly modeling probabilities and the fact that logistic regression is less sensitive against outliers (Henking et al., 2006). For these reasons, the logistic regression method is presented as basis for the classical way of credit scorecard development.
The logit model represents a generalized linear model where the probit model or the complementary log–log model are also representatives (Fahrmeir et al., 2013). The aim of the regression analysis for the binary outcome variable Y_i (default and non-default) is the analysis of the probability $\pi_i = P(Y_i = 1 \mid x_i) = G(x_i'\beta)$ with x_i as a vector of explanatory variables and $x_i'\beta = \beta_0 + \beta_1 x_1 + \ldots + \beta_p x_p$ as linear predictor. The logistic regression model supposes a linear relationship between the log-odds $ln\frac{\pi_i}{1-\pi_i}$ and the covariates. The logit model arises by choosing the following response function:

$$G(t) = (1 + \exp(-t))^{-1} \tag{3.1}$$

The maximum likelihood estimation is usually used as a method to estimate the β-coefficients. The rationale of this method describes the maximization of the likelihood $L(\beta)$ with the definition:

$$L(\beta) = \prod_{i=1}^{n} G(x_i'\beta)^{Y_i}(1 - G(x_i'\beta))^{1-Y_i} \tag{3.2}$$

The maximum likelihood estimator $\hat{\beta}_{ML}$ results from maximizing the likelihood $L(\beta)$ and the log-likelihood $l(\beta) = log(L(\beta))$, respectively.
The theory of the maximum likelihood estimation denotes the following properties: for $n \to \infty$ the ML-estimator asymptotically exists, is consistent and asymptotically normal distributed.

This implies for a sufficiently large sample size n, the approximately normal distribution for $\hat{\beta}$

$$\hat{\beta}_{ML} \rightarrow N(\beta, F^{-1}(\beta)) \tag{3.3}$$

The asymptotic variance matrix equals the inverse Fisher information and the variance matrix of the weighted regression with $Var(Y) = D(\beta) = diag(G(x_i'\beta)(1 - G(x_i'\beta)))$. For further explanations, I refer to Fahrmeir et al. (2009) and Hosmer and Lemeshow (2000). For testing the significance of the coefficients in the logit model, the Wald test and the likelihood quotient test are normally presented.
However, testing the significance is not the only criterion for selecting explanatory variables. Finding the best covariates for the final model is an important aspect but the topic of variable selection goes beyond the scope of this thesis. Scheipl et al. (2012) and Fahrmeir et al. (2013) are referenced for further reading.

3.2 The Development of Scorecards

The logit model is the most commonly applied method in practice to create scoring models within the scorecard development process. Therefore, the empirical results from the logistic regression are presented and used as a benchmark for the new methods from statistics and machine learning. For the credit scoring procedure, various steps and analyses are needed for selecting and estimating the final model. In this section, the main issues are highlighted for developing a credit scoring model (cf. for instance Anderson (2007)).

The data and the definition of the criterion of default are described in Section 2.2. The univariate analysis is one of the most important parts of the scorecard development process. All potential covariates are considered within these investigations. This includes, among other surveys, the analysis of the distribution and discriminative power for each attribute, the plausibility and relevance of the variables and the consideration of missing values and outliers. The latter aspects are negligible in the analysis because of the excellent data quality. Missing values are, therefore, no problem in the presented credit scoring case. Moreover, the few identified outliers do not change the results. This is also verified in the multivariate context described later.

One example of the univariate analysis, Figure 3.1, shows the monthly univariate AUC values of variable 1. Three different criteria of default are displayed in the figure. The aim is to test the stability of the variables over time. The displayed variable shows a stable distribution for the three definitions of a default since the jumps arise due to small amounts of observations in a month. If the distribution would imply instability, the reasons have to be investigated to decide whether the corresponding covariate should be included in the scoring model or not.

Only variables with stable risk development and distribution in the past are used in order to create a robust model for the future decision process of the retail bank.
Another aspect within the univariate analysis is the classification of continuous variables in order to use them with categorized classes. Typically, in retail credit scoring, numerical

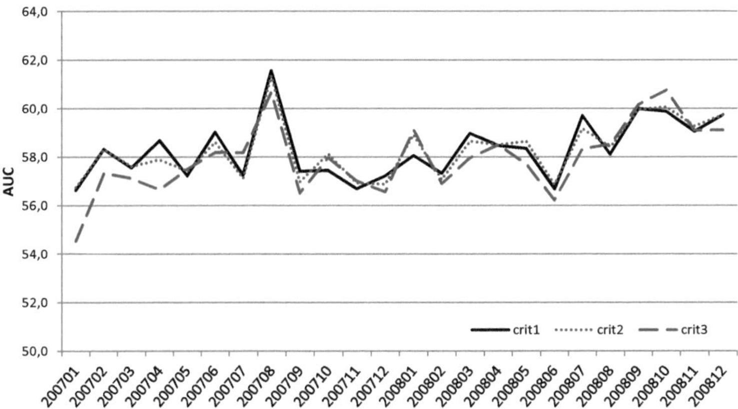

Figure 3.1: Monthly univariate AUC values of variable 1 for three different criterions of default (crit1 to crit3) to test the stability of the variable over time.

covariates are grouped into intervals. An advantage is that all variables are of the same kind. Another advantage is the interpretability (Hand, 2001). Different groups can easily be characterized as more or less risky compared to other groups. In the current case, the variables are categorized regarding the default rate in an interval, the amount of observations and the amount of defaults in a class. The univariate predictive performance of the covariate is measured by the AUC. Hand and Adams (2000) discuss the properties of variable classification. Simple explanation and the use of the same variable types are already mentioned as attractive properties. Another advantage is the modeling of nonlinear relationships between the raw score and the weights, and therefore between the raw scores and the outcome variable (Hand and Adams, 2000). A main weakness, however, is the loss of information by transforming a variable and treating neighboring values identically in a scoring model. Additionally, the jumps in weights might be unrealistic by crossing a threshold.

The classification of covariates is not necessary for many statistical predictive models. In this thesis, categorized variables and continuous covariates are considered depending on the applied models. As described in Section 2.2, four of the variables are categorical. Whether the covariates are classified or not always refers to the 22 numerical explanatory variables.

Finally, the univariate analysis results in a so-called short list. This contains the selected attributes as basis for the multivariate analysis and a ranking order of the importance of predictive performance as measured by the AUC.

Figure 3.2 displays the 26 potential covariates and shows the ranking order concerning the univariate AUC measure. Due to confidentiality reasons, the covariates can not be published in detail. But regarding further analyses with new statistical methods, I highlight the importance of variables 3 and 4. These are the best covariates concerning the presented criterion of default and the univariate predictive performance. Variables 10 and 20 are of minor importance. Note that variables 5, 6, 8 and 9 are original categorical variables, while the other 22 variables are classified.

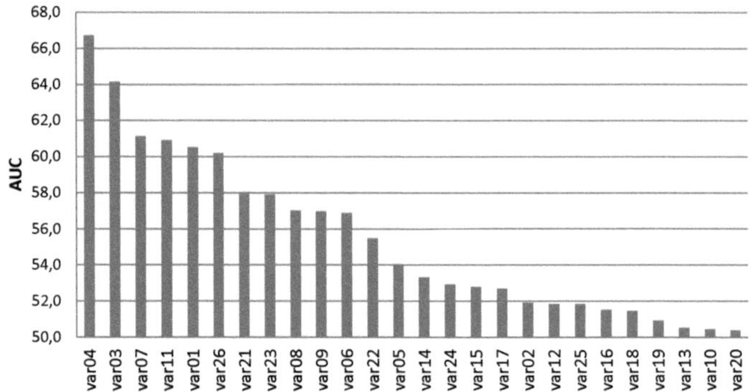

Figure 3.2: Ranking order of the 26 classified potential covariates according to the univariate AUC values analyzed within the classical scorecard development process.

The aim of the multivariate analysis is to find the optimal combination of scoring variables for the final scoring model. Hundreds of different models are tested regarding correlations, significance, plausibility and the AUC as a measure of performance. The coefficients are estimated with the maximum likelihood method and tested for significance. As previously mentioned in Section 3.1, variable selection is not the main focus in this thesis. The final and optimal model with respect to AUC includes 8 explanatory variables. In addition to the AUC, another criterion is considered for the variable selection in order to determine if this would lead to different results. Considering the model selection with the Akaike information criterion (AIC) of Akaike (1981), the results for the presented credit scoring case are similar and lead to the same final scoring model. For the final results, different performance measures were evaluated.

For the final model with 8 variables, backtesting analyses were conducted to prove the stability and robustness of the multivariate model. Time series for the predictive power, plots of the distribution of good and bad credits, or a graphical illustration for the default rate over the scores are possible investigations.

In Table 3.1, the results for the final credit scoring model are shown that were estimated by logistic regression with coarse classification.

		AUC	Gini	H measure	KS	MER	Brier score
trainORG	train	0.7558	0.5118	0.1943	0.3881	0.0258	0.0255
	test	0.7204	0.4408	0.1429	0.3232	0.0262	0.0262
	validation	0.7263	0.4527	0.1468	0.3310	0.0260	0.0260
train020	train	0.7610	0.5219	0.2226	0.3973	0.1929	0.1378
	test	0.7211	0.4423	0.1421	0.3255	0.0263	0.0637
	validation	0.7282	0.4564	0.1476	0.3314	0.0260	0.0658
train050	train	0.7570	0.5140	0.2324	0.3860	0.3070	0.2019
	test	0.7211	0.4423	0.1421	0.3255	0.0263	0.2121
	validation	0.7275	0.4550	0.1473	0.3364	0.0260	0.2217

Table 3.1: Logistic regression results for the original training sample, training samples with higher default rates, and application to the test and validation samples with classified variables.

For the training samples, the AUC values are, as expected, quite high, with values of 0.7560 for the original training sample. The corresponding Gini coefficients for the estimated model vary around 51% for the training samples, while the values for the test sample with the applied model, are closely spaced with an AUC measure around 72% and Gini coefficients around 44%. The H measure, KS, MER and the Brier score are also shown in Table 3.1. For instance, the H measure on the validation data denotes 0.1468, 0.1476 and 0.1473 (trained on trainORG, train020 and train050).

For the variable selection, the original training sample is used with the default rate of 2.6%. For the estimation of the simulated training samples with default rates of 20% and 50%, the selection of the explanatory variables is maintained and the coefficients are estimated on the new samples. The predictive accuracy for the samples with different default rates differs only slightly. The results for the different training samples are quite robust.

The presented outcomes for the classical scorecard, estimated with the logistic regression model in Table 3.1, result from models with classified covariates. The models with continuous explanatory variables yield lower prediction accuracies than the models with categorized variables. The results of the logit model with continuous covariates are also denoted in this thesis for comparison reasons. For the analyses, it is always mentioned if coarse classification is used or continuous variables are employed.

For the models with continuous covariates, the models are also analyzed with variables where the values above the 95% quantile and under the 1% quantile are set to the quantile value. This was done to analyze the influence of outliers. The results for these models are approximately the same as the results for the former models. This confirms the findings of the univariate analysis that there are no severe outliers in this credit scoring data.

In credit scoring, the estimation of the multivariate model is typically transformed to so-called score values. The estimations are therefore converted to a specific scale. An example would be a range of 1 - 1000, where the probability estimates of each customer are multiplied with 1000 and rounded out (Henking et al., 2006). These scores are the basis for the calibration of the scoring model. The calibration implies the mapping of the score values to rating classes. Score ranges are built on the basis of rating classes with pre-specified default rates. Therefore, a lot of different secondary conditions have to be satisfied for the calibration. These rating classes can then be used for the definition of the cut-off to define, which customers should be accepted and rejected, respectively. Furthermore, the rating classes can be used for the default risk forecasts (Thomas, 2009).

Another difficult aspect for building application scorecards is that no default information exists about rejected applicants. Various techniques are proposed for the so-called reject inference (Thomas, 2009): Reclassification and reweighting are two of the methods used for handling reject cases. The first method considers applications as defaults when they have negative characteristics leading always to a reject application. Reweighting implies the fact that the defaults and non-defaults in a special range are weighted with the part of the rejected cases in that range. Further details and more reject inference methods are described in Thomas (2009). In the literature, an active discussion arose on methods to include rejected cases in the design sample. Compare to the research on reject inference for example Hand and Henley (1993), Hand and Henley (1997), Crook and Banasik (2004) and Banasik and Crook (2005). In the meantime, most surveys conclude that there is no better performance by using reject inference methods. A reasonable strategy would be to randomly select and accept applicants who would normally be rejected and get information about their default behavior. In the banking area, this strategy is seldom applied because it is costly.

Since the application of reject inference methods is critical, all data samples, that are selected from the whole portfolio and presented in this thesis, are based on accepted credit customers, where the default information could be observed.

3.3 Discussion of the Logistic Regression Model in Credit Scoring

Good interpretability, simple explanation and robust models are three of the reasons for the popularity of the logistic regression models in credit scoring. Another aspect is the well established theoretical background of the regression framework. Generalized linear models are implemented in every standard statistical software package. The computation time of the models is efficient even for very large data sets. This makes it possible to estimate hundreds of different models in an acceptable time period.

The logistic regression model, however, assumes a logistic relationship. It is limited to the additive form of the model and assumes linearity in the covariates. In this thesis, different methods are proposed to overcome these aspects in the credit scoring context. For instance, the AUC approach in Chapter 4 overcomes the underlying distribution assumption of the logit model, while the generalized additive models in Chapter 5 include nonlinearities. The machine learning algorithms offer non-parametric approaches.

The reason to use the classical logistic regression model as benchmark for the further presented evaluation is that the logit model is still the most commonly applied method for the scorecard development process in the banking sector. Other algorithms are normally not implemented and used for estimating credit scorecards. Therefore, the comparison is always drawn to the logit model and highlight the advantages and disadvantages to the mentioned one. Additionally, some insights and comparisons are given of the different methods among each other. The focus in this thesis is not on the variable selection of the covariates.

In the context of the scorecard development with logistic regression, even if the interpretability and the implementation of the method is feasible, the whole development process from the data preparation to the final model, needs a lot of time for investigations.

In general, it is worth evaluating newer algorithms to improve the model building since the optimization of the selling process of the bank and the optimization of the risk prediction offers great opportunity for economization.

Chapter 4

Optimization AUC

4.1 A New Approach for AUC Optimization

The performance measure AUC and the Gini coefficient are popular indicators for evaluating predictive accuracy in credit scoring. Classification algorithms, however, are not necessarily optimal with respect to these measures. The aim is to use the AUC not only as a tool for comparing the results, but also to introduce it as an objective function and apply the approach to the data from the consumer credit market.

In the scientific field of biostatistics, Pepe et al. (2006) propose the AUC as an objective function with application to protein biomarkers. They use the Mann–Whitney statistic for the nonparametric AUC estimation. Pepe et al. (2006) receive superior results with the AUC-based linear combination compared to the logistic regression model, particularly when the model assumption failed. However, the authors remark that additional work needs to be done before widely applying this approach. Among other reasons, only two markers are used for the application (Pepe et al., 2006). Ma and Huang (2005) extend the approach with the sigmoid approximation for the AUC for high-dimensional data. Using the ROC technique for microarray data, they present a gradient descent technique for their analyses. For the credit scoring context, gradient descent methods are considered within the boosting framework in Section 7. Eguchi and Copas (2002) also cover the topic of AUC optimization within linear scores by dealing with a complex calculating method for the AUC. Moreover, they only present an example of limited breast cancer data to show the improvements over the logit model.

Miura et al. (2010) propose a sigmoid-approximated AUC maximization method and attempt credit scoring for companies from the Tokyo Stock Exchange. For the calculation of the AUC, they follow the definition of Mann–Whitney and an approximation with a sigmoid function. They obtain good results regarding predictive power, especially while considering outliers. However, they employ a small sample of 75 companies and use three explanatory variables.

The emphasis is to introduce the AUC as the objective function for β-estimation, calculate the measure with the well-known Wilcoxon–Mann–Whitney statistic (Wilcoxon,

1945; Mann and Whitney, 1947), use the Nelder–Mead algorithm for optimizing the coefficients, and apply the approach to the retail credit scoring data. Extending the definition of equation 2.4, β^t is introduced as a vector of coefficients, while x_d and x_{nd} denote the scores as vectors of explanatory variables:

$$AUC(\beta) = \frac{1}{n_d \cdot n_{nd}} \sum_1^{n_d} \sum_1^{n_{nd}} S(\beta^t(x_d - x_{nd})) \tag{4.1}$$

The aim is to optimize the β-coefficients by maximizing $AUC(\beta)$:

$$\hat{\beta}_{AUC} = \arg\max_{\beta} AUC(\beta) \tag{4.2}$$

$$= \arg\max_{\beta} \frac{1}{n_d \cdot n_{nd}} \sum_1^{n_d} \sum_1^{n_{nd}} S(\beta^t(x_d - x_{nd})) \tag{4.3}$$

$AUC(\beta)$ is a sum of step functions, and therefore, it is not continuous and not differentiable in terms of β. The proposed method is an unconstrained optimization problem. The difference of the scores is estimated using $\beta_1 \mathrm{v1}_d - \beta_1 \mathrm{v1}_{nd} + \beta_2 \mathrm{v2}_d - \beta_2 \mathrm{v2}_{nd} + \ldots + \beta_t \mathrm{vt}_d - \beta_t \mathrm{vt}_{nd}$, with t and v as the number of variables and the explanatory variables, respectively.

For optimizing the coefficients, the derivative-free method of Nelder and Mead (1965) is used here. The algorithm is a direct search method and one of the best known algorithms for unconstrained optimization of nonsmooth functions. Let $f : \mathbb{R}^n \to \mathbb{R}$ be a nondifferentiable function without information about the Gradient $\nabla f(x)$ and the Hessian matrix $\nabla^2 f(x)$. The rationale to find the minimum of f is to generate a sequence of simplices that should realize a closer diameter at each iteration, so as to finally determine the minimum (Geiger and Kanzow, 1999). The algorithm that is used here describes a simplex method, which differs from the more famous simplex algorithm for linear programming. For a detailed description of the procedure and more references, see Lagarias et al. (1998).

While linear programming can also be used for developing scoring models, it denotes, in contrast to the presented AUC approach, a minimization problem with constraints. A description for the credit scoring case is given in Thomas (2000), which denotes the minimization of the sum of misclassification errors. Furthermore, Thomas (2009) proposes an approach for maximizing divergence, which is also used in the credit scoring area. The divergence was introduced by Kullback and Leibler (1951) measuring the difference between two probability distributions. The aim of the method for building a scorecard with coarse classified attributes is to find an additive score, which maximizes divergence. This approach represents a nonlinear maximization problem, where iterative procedures are used to improve the divergence from one iteration to the next (Thomas, 2009). The AUC approach is also compared to the divergence approach for variables with coarse classification.

4.2 Optimality Properties of the AUC Approach

4.2.1 Theoretical Considerations

Theoretical considerations are given in the following for the proposed AUC approach explained in Section 4.1. The theoretical properties of the AUC optimization can be considered as in Pepe et al. (2006). They explore the linear scores of the form $L_\beta(x_i) = x_1 + \beta_2 x_2 + \ldots + \beta_P x_P$ without intercept and 1 for the coefficient of x_1. Rules based on $L_\beta(x_i) > c$, with c as a threshold $c \in (-\infty, \infty)$, are optimal in accordance with the Neyman–Pearson lemma (Neyman and Pearson, 1933) if the score is a monotone increasing function of $L_\beta(x_i)$

$$P[Y_i = 1 \mid x_i] = g(x_1 + \beta_2 x_2 + \ldots + \beta_P x_P) = g(L_\beta(x_i)) \tag{4.4}$$

This implies that there is no other classification rule based on x_i lying above the ROC curve for $L_\beta(x_i)$. Pepe et al. (2006) explain that if the false positive rate (FPR) is fixed, there is no true positive rate (TPR) for any other classification with the same FPR higher than the TPR for the rule $L_\beta(x_i) > c$. Conversely, if the TPR is fixed, there is no rule based on x_i lower than the rule $L_\beta(x_i)$. As a consequence of the optimality of the ROC curve for $L_\beta(x_i)$, Pepe (2003) shows that the overall misclassification rate for rules with $L_\beta(x_i) > c$ is minimized, and the cost of false-negative and false-positive errors is minimized.

Pepe et al. (2006) point out that $L_\beta(x_i)$ has the best ROC curve for all linear predictors $L_b(x_i) = x_1 + b_2 x_2 + \ldots + b_P x_P$, assuming that $L_\beta(x_i)$ has the best ROC curve among all rules based on x_i. The aim is to find coefficients for $(b_2, \ldots, b_P)$ with the best empirical ROC curve. Since the optimal ROC curve has the maximum AUC value, this measure can be used as an objective function to estimate β.

While the Neyman–Pearson lemma (Neyman and Pearson, 1933) was developed for statistical hypothesis testing, it also applies to any decision problem. While Pepe (2003) considers the analogy between statistical hypothesis testing and medical diagnostic testing, it applies equally well to the credit scoring problem. Table 4.1 illustrates the analogy.

The possible states in statistical hypothesis testing H_0 and H_1 are the events of default and non-default in the credit scoring case. The false positive fraction represents the significance level in hypothesis testing, while the true positive fraction describes the statistical power. Considering the likelihood ratio function $LR(W)$, the Neyman–Pearson lemma can be described for statistical hypothesis testing as follows (Pepe, 2003). The form of the decision rule maximizing the statistical power among all other decision rules based on W with the significance level α, has the form $LR(W) > c$. The threshold c is chosen in the way that $\alpha = P[LR(W) > c \mid H_0]$. The LR function is assumed to be monotone increasing according X. For the credit scoring problem, the Neyman–Pearson lemma is previously stated for the linear scores $L_\beta(x_i)$ and the ROC curve with the assumption that the score is a monotone increasing function of $L_\beta(x_i)$.

To cover the convergence properties for the best empirical ROC curve and the β-estimation with the AUC as an objective function, it is important to consider the relationship

	Statistical hypothesis testing	Credit scoring problem
Possible states	H_0 versus H_1	$Y = 0$ versus $Y = 1$
Information	Data for n subjects denoted by W	Test result for a subject denoted by X
Rule	Classify as H_0 or H_1 using W	Classify as $Y = 0$ or 1 using X
Type 1 error	Significance level $\alpha = P[\text{reject } H_0 \,\vert\, H_0]$	False positive fraction $FPF = P[\text{classify } Y = 1 \,\vert\, Y = 0]$
Type 2 error	Power $1 - \beta = P[\text{reject } H_0 \,\vert\, H_1]$	True positive fraction $TPF = P[\text{classify } Y = 1 \,\vert\, Y = 1]$
LR function	$LR(W) = P[W \,\vert\, H_1]/P[W \,\vert\, H_0]$	$LR(X) = P[X \,\vert\, Y = 1]/P[X \,\vert\, Y = 0]$

Table 4.1: Analogy of the Neyman–Pearson lemma between statistical hypothesis testing and the credit scoring problem.

to the maximum rank correlation (MRC) estimator of β from Han (1987) (Pepe et al., 2006). The relation to the Mann–Whitney statistic is already mentioned in Section 2.1 and 4.1.

Han (1987) considers the generalized model

$$y_i = D \cdot F(x_i'\beta_0, u_i) \tag{4.5}$$

with $i = 1, \ldots, N$, u_i as an iid error term, the assumption for $D \cdot F$ that $D : \mathbb{R} \to \mathbb{R}$ is non-degenerate monotonic and $F : \mathbb{R}^2 \to \mathbb{R}$ is strictly monotonic. The binary choice model is a special case with $y_i = D \cdot F(x_i'\beta_0, u_i) = D(x_i'\beta_0 + u_i)$ for

$$D(\xi) = 1 \quad \text{if} \quad \xi \geq 0 \tag{4.6}$$

$$= 0 \quad \text{otherwise} \tag{4.7}$$

Maximizing the rank correlation between the y_i and the $x_i'\beta$ with respect to β is the idea of the maximum rank correlation estimator. The rationale is that given an inequality $x_i'\beta_0 > x_j'\beta_0$ for a pair of samples, it is likely that $y_i > y_j$. This implies a positive correlation between the ranking of the y_i and the ranking of the $x_i'\beta_0$. Consider the following

$$S_N(\beta) = \begin{bmatrix} N \\ 2 \end{bmatrix}^{-1} \sum_{\rho} \left[l(y_i > y_j) l(x_i'\beta > x_j'\beta) + l(y_i < y_j) l(x_i'\beta < x_j'\beta) \right] \tag{4.8}$$

with $l(\cdot)$ as indicator function, $l(\cdot) = 0$ if $(\cdot)$ is true, $l(\cdot) = 1$ otherwise and $\sum_\rho$ as summation over $\binom{N}{2}$ combination of two distinct elements (i, j). $S_N(\beta)$ denotes the rank correlation between y_i and the $x_i'\beta$ considering only their rankings (Han, 1987). The definition of the estimator $\hat{\beta}$ of the parameters β_0 results by maximizing $S_N(\beta)$

$$S_N(\hat{\beta}) = \max_{\beta} S_N(\beta) \tag{4.9}$$

The maximum rank correlation (MRC) estimator $\hat{\beta}$ shows scale invariance in the estimation of β_0 following $S_N(\beta) = S_N(\gamma\beta)$ for any constant $\gamma > 0$. The proof that the MRC estimator is $\sqrt{n}$-consistent and asymptotically normal is given in Sherman (1993). The proof is based on a general method for determining the limiting distribution of a maximization estimator which requires neither differentiability nor continuity of the criterion function. Moreover, the proof is based on a uniform bound for degenerate U-processes (Sherman, 1993). Since at least one component of x_i' is continuous, the estimator is consistent and asymptotically normal under the generalized linear model 4.4 (Sherman, 1993). Sherman (1993) is referenced for further details.

4.2.2 Simulation Study

To compare the AUC approach and the logistic regression model, a simulation study is conducted. The AUC approach is a distribution free approach that does not rely on a special link function. Compared to the logit model, the AUC approach should be quite robust even if the link function is not logistic. While the logit model assumes a logistic link function, the AUC approach only requires a monotone function.

For the simulation study, the logit model and the AUC approach are compared by simulating data with the logit, probit and complementary log–log link function. Three normal distributed explanatory variables are created with $\mu = 0$ and $\sigma^2 = 1$. The binary response variable is generated from a Bernoulli distribution $Y \sim Ber(\pi)$, where π is determined using the specific link function.

The response function for the logistic model is already stated in equation 3.1 (Section 3.1) for analysing the probability $\pi_i = P(Y_i = 1 \,|\, x_i) = G(x_i'\beta)$, where x_i denotes a vector of explanatory variables and $x_i'\beta = \beta_0 + \beta_1 x_1 + \ldots + \beta_p x_p$ describes the linear predictor. For simulating the probit case, the distribution function Φ from the standard normal distribution is used as follows

$$G(t) = \Phi(t) = \Phi(x_i'\beta) \tag{4.10}$$

with the related link function $\Phi^{-1}(\pi_i) = x_i'\beta = t$.
To simulate the complementary log–log link in the data, the link function $\log(-\log(1-\pi_i)) = x_i'\beta = t$ is used from the complementary log–log model with the extreme value distribution and the following response function:

$$G(t) = 1 - \exp(-\exp(t)) \tag{4.11}$$

Figure 4.1 shows the logit (3.1), probit (4.10) and complementary log–log (4.11) response functions. Since the latter function is asymmetrical, the response functions of the logit and probit model are both symmetrically around zero. Considering the linear predictor towards infinity, the response function of the probit model tends faster to 0 and 1 respectively compared to the logistic distribution function. For $x_i'\beta \to \infty$, the response function of the complementary log–log model tends faster to 1 than the two other distributions.

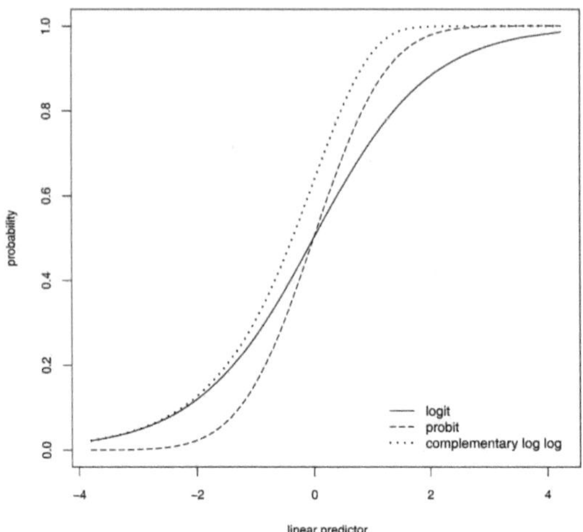

Figure 4.1: Response functions for the logit, probit and complementary log log model.

Results for statistical analysis with the logit and probit model are quite similar concerning the estimated probabilities. In Fahrmeir et al. (2009), it is stated that the estimated coefficients of a logit model differ approximately with the factor $\sigma = \pi/\sqrt{3} \approx 1.814$ from the coefficients of a probit model (with $\sigma^2 = 1$), while the estimated probabilities are quite similar. Compared to the logit model, the extreme value distribution of the complementary log–log model with the variance $\sigma^2 = \pi^2/6$ and an expected value of -0.5772 has to be adjusted to a response function with the variance $\sigma^2 = \pi^2/3$ and an expected value of 0 (Fahrmeir et al., 2009). This implies that the results for statistical analysis with the complementary log-log model differ more explicitly from the logit and probit model.

To analyze link functions, which differ from the abovementioned functions, the logit link and the complementary log–log link are adjusted. Based on the logit link, a function (link1) was created, which represents the logit link for $t \leq 0$ and tends faster to 1 for $t > 0$

$$H(t) = \begin{cases} G(t) & \text{for } t \leq 0 \\ G(8 * t) & \text{for } t > 0 \end{cases} \tag{4.12}$$

with $G(t) = (1 + \exp(-t))^{-1}$ as the logit link. Figure 4.2(a) shows the function compared to the logistic response function. Figure 4.2(b) shows another simulated function based on the complementary log–log model. Since the curve tends slower to zero for $x_i'\beta \to -\infty$, the function tends much faster to 1 than the complementary log–log and the logit response functions. For comparison reasons, the logit link is also contained in Figure 4.2(b). The following equation represents the link function (link2)

$$Q(t) = \begin{cases} G(0.2 * t) & \text{for } t \leq 0 \\ G(5 * t) & \text{for } t > 0 \end{cases} \tag{4.13}$$

with $G(t) = 1 - \exp(-\exp(t))$ as complementary log–log link function.

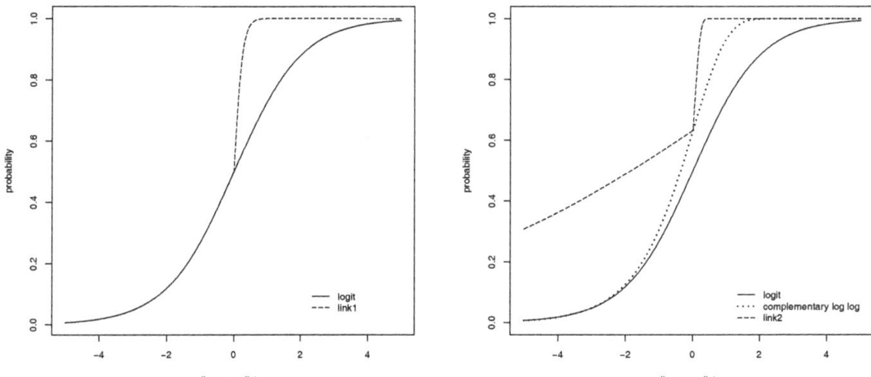

(a) Modified link function based on the logit link (b) Link function based on the compl. log log link

Figure 4.2: Two adjusted link functions for the simulation study to analyze the properties of the AUC approach and the logit model.

The two simulated functions differing from the logit, probit and complementary log–log link, should also illustrate the properties of the AUC approach and the logistic regression. For the data generating process, the coefficients that were used to produce the three explanatory variables, are

$$\beta = (1, 0.5, 0.3)$$

Thus, the simulated data contain a binary response and three independent normally distributed variables produced with the coefficients and a specified link function. 1000 observations are produced for each data set ($n = 1000$) and 100 simulations are run in each step. In addition to the simulated training data, validation data are simulated to validate the outcomes. The simulation procedure can be described as follows; simulating training data as described above, estimating the coefficients with the logit model and the AUC approach, calculating the AUC measure for the two approaches on the training data, simulating validation data, applying the estimated coefficients of the two different estimation procedures to the validation data, calculating the AUC values for both approaches on the validation data and finally running the procedure 100 times.

Note that the coefficients estimated with the maximum likelihood method are used as starting values for the AUC optimization. In the simulation study, both starting values for the coefficients of $x2$ and $x3$ are divided by the coefficient of $x1$. The first coefficient parameter is fixed to 1 because of normalization reasons.

The results for the estimated coefficients of the variables $x2$ and $x3$ are presented first. Figure 4.3 contains the estimated coefficients for the two variables and compares the values for the logistic regression and the AUC approach assuming specific link functions. In addition, Table 4.2 shows the mean values and the standard deviations for the estimated coefficients on the training data for both methods.

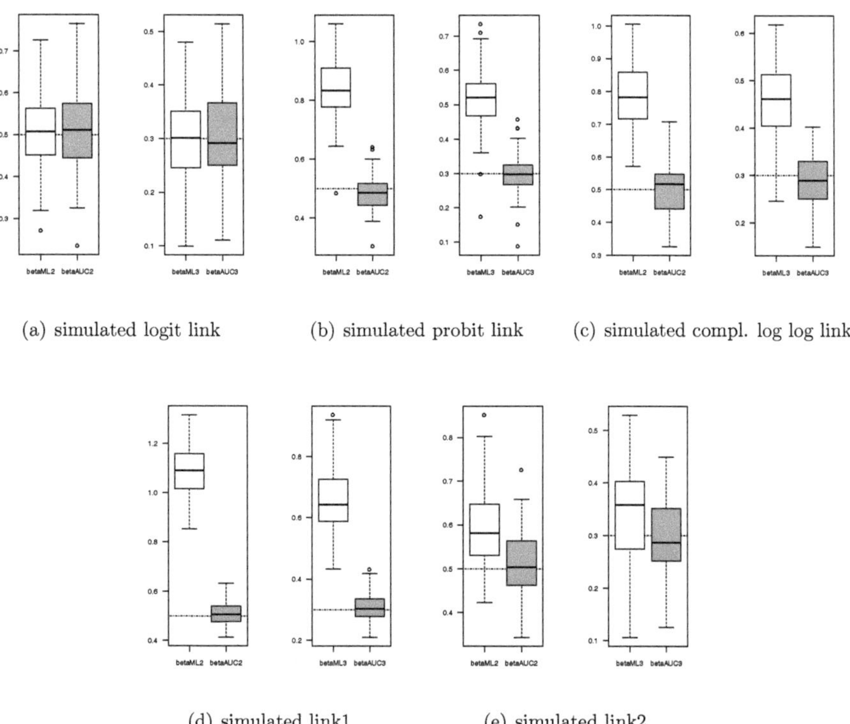

(a) simulated logit link (b) simulated probit link (c) simulated compl. log log link

(d) simulated link1 (e) simulated link2

Figure 4.3: Comparison of the estimated coefficients for variable $x2$ (left) and $x3$ (right) for the maximum likelihood estimation (white) and the AUC approach (grey) on the simulated training data with five different link functions.

As described in Section 4.2.1, $\hat{\beta}_{ML}$ and $\hat{\beta}_{AUC}$ are both consistent estimates of β, and both approaches yield the optimal linear combination.
Regarding the coefficient $\hat{\beta}_2$ and $\hat{\beta}_3$ for the simulated logit link, it is indicated that the estimates of the maximum likelihood estimation are more efficient than the estimates of the AUC approach. The standard deviations denote smaller values for $\hat{\beta}_{ML_2}$ and $\hat{\beta}_{ML_3}$ than for $\hat{\beta}_{AUC_2}$ and $\hat{\beta}_{AUC_3}$. For all other simulated link functions, the standard deviations for the AUC estimates are smaller than the ones for the maximum likelihood estimates. Moreover, the mean values for the AUC estimates are close to the values of 0.5 and 0.3 for β_2 and β_3, respectively. The box plots in Figure 4.3 show this effect. Figure 4.3(a) denotes similar mean values of the maximum likelihood estimates and the AUC estimates for the simulated logit link. The other boxplots in Figure 4.3(b) to Figure 4.3(e) show for the means of the coefficients of both variables, that the AUC estimates are closer to the simulated values in contrast to the maximum likelihood estimates. A visualization of the coefficients of two explanatory variables and the corresponding AUC measures is shown in Figure 4.5 in Section 4.3.

	$\hat{\beta}_{ML_2}$		$\hat{\beta}_{AUC_2}$		$\hat{\beta}_{ML_3}$		$\hat{\beta}_{AUC_3}$	
	mean	sd	mean	sd	mean	sd	mean	sd
logit	0.508787	0.085478	0.510491	0.100341	0.302441	0.074741	0.305843	0.079996
probit	0.839319	0.104409	0.485139	0.057669	0.517261	0.088749	0.299602	0.058298
comp log log	0.783629	0.093954	0.500100	0.069781	0.456872	0.078068	0.292255	0.055788
link1	1.090074	0.102941	0.509633	0.044181	0.654038	0.102251	0.308601	0.046670
link2	0.589547	0.082155	0.509457	0.069216	0.343642	0.091897	0.294575	0.069156

Table 4.2: Means and standard deviations for the coefficients of variables $x2$ and $x3$ for the simulated training data with various link functions, estimated with the logit model ($\hat{\beta}_{ML}$) and the AUC approach ($\hat{\beta}_{AUC}$).

Besides considering the estimated coefficients, Table 4.3 contains the results of the classification accuracy for 100 simulations. The mean values and standard deviations of the AUC measure and other measures are presented for the simulated data and specific link functions. On the training data, the predictive accuracy of the AUC approach always outperforms the accuracy of the logistic regression, regardless of the simulated link function in the data. The models trained on the training data are applied to the validation data. These outcomes are of great importance. Assuming the logit link in the data, the mean value of the AUC for the AUC approach on the validation data is slightly below the predictive accuracy of the logit model. However, the results show only a tendency since the AUC values are quite close to each other. For the probit link as well as for the complementary log–log link in the simulated data, the predictive accuracy of the AUC approach and the logit model equals on the validation data. The mean value of the data with probit link have an AUC value of 0.8589 for both methods.

Two link functions are created, which explicitly differ from the formerly mentioned functions. Simulating the link1 function in the data, the predictive accuracy for the AUC approach is the same as the corresponding AUC value for the logit model. The main difference between the two approaches is demonstrated with the link function that differs mostly from the logit one. The mean AUC for the simulated validation data with link2 is 0.8202 compared to 0.8209 for the AUC approach (cf. Table 4.3). The Brier score for link2 on the validation data indicates that the predictive performance of the logit model is better than that of the AUC approach. The other performance measures, however, indicate the same tendency as the AUC. In the table, cases are indicated with (1) and (2) where the AUC approach outperforms, or is equivalent to the logit model, and where the logit model provides better results, respectively.

		training				validation			
		logReg		AUC approach		logReg		AUC approach	
		mean	sd	mean	sd	mean	sd	mean	sd
logit	AUC	0.7676	0.0153	0.7679$^{(1)}$	0.0153	0.7643$^{(2)}$	0.0146	0.7642	0.0144
	H	0.2613	0.0251	0.2620$^{(1)}$	0.0256	0.2557$^{(2)}$	0.0240	0.2553	0.0238
	KS	0.4101	0.0269	0.4111$^{(1)}$	0.0271	0.4062	0.0270	0.4062$^{(1)}$	0.0266
	MER	0.2942	0.0134	0.2936$^{(1)}$	0.0137	0.2964	0.0134	0.2964$^{(1)}$	0.0132
	Brier	0.1965$^{(2)}$	0.0061	0.1969	0.0061	0.1986	0.0058	0.1985$^{(1)}$	0.0059
probit	AUC	0.8597	0.0116	0.8599$^{(1)}$	0.0116	0.8589	0.0104	0.8589$^{(1)}$	0.0104
	H	0.4340	0.0259	0.4347$^{(1)}$	0.0261	0.4328$^{(2)}$	0.0231	0.4325	0.0230
	KS	0.5598	0.0281	0.5605$^{(1)}$	0.0281	0.5612$^{(2)}$	0.0225	0.5609	0.0224
	MER	0.2197	0.0141	0.2195$^{(1)}$	0.0142	0.2192	0.0112	0.2191$^{(1)}$	0.0112
	Brier	0.1532$^{(2)}$	0.0063	0.1611	0.0050	0.1543$^{(2)}$	0.0060	0.1613	0.0044
comp log log	AUC	0.8501	0.0126	0.8504$^{(1)}$	0.0126	0.8480	0.0116	0.8480$^{(1)}$	0.0116
	H	0.4018	0.0276	0.4023$^{(1)}$	0.0274	0.3958$^{(2)}$	0.0250	0.3953	0.0246
	KS	0.5503	0.0249	0.5517$^{(1)}$	0.0242	0.5438$^{(2)}$	0.0245	0.5433	0.0243
	MER	0.2233	0.0137	0.2232$^{(1)}$	0.0131	0.2258	0.0134	0.2258$^{(1)}$	0.0131
	Brier	0.1547$^{(2)}$	0.0068	0.1718	0.0055	0.1562$^{(2)}$	0.0067	0.1727	0.0050
link1	AUC	0.9027	0.0096	0.9029$^{(1)}$	0.0095	0.9009	0.0101	0.9009$^{(1)}$	0.0101
	H	0.5467	0.0277	0.5475$^{(1)}$	0.0276	0.5421	0.0262	0.5425$^{(1)}$	0.0263
	KS	0.6932	0.0231	0.6940$^{(1)}$	0.0226	0.6896	0.0188	0.6901$^{(1)}$	0.0192
	MER	0.1740	0.0124	0.1736$^{(1)}$	0.0123	0.1757	0.0119	0.1755$^{(1)}$	0.0118
	Brier	0.1274$^{(2)}$	0.0065	0.1542	0.0050	0.1294$^{(2)}$	0.0070	0.1549	0.0050
link2	AUC	0.8215	0.0125	0.8222$^{(1)}$	0.0124	0.8202	0.0117	0.8209$^{(1)}$	0.0116
	H	0.3070	0.0228	0.3101$^{(1)}$	0.0225	0.3052	0.0196	0.3083$^{(1)}$	0.0189
	KS	0.5772	0.0218	0.5831$^{(1)}$	0.0204	0.5731	0.0194	0.5775$^{(1)}$	0.0186
	MER	0.2189$^{(2)}$	0.0135	0.2193	0.0133	0.2213	0.0127	0.2212$^{(1)}$	0.0129
	Brier	0.1455$^{(2)}$	0.0065	0.2138	0.0067	0.1474$^{(2)}$	0.0070	0.2132	0.0065

Table 4.3: Means and standard deviations of different performance measures for the logit model and the AUC approach estimated for the simulated training samples and applied to simulated validation data with different link functions. (1) denotes cases where the AUC approach outperforms or is equivalent to the logit model. (2) denotes cases where the logit model provides better results.

While the different performance measures are indicated in Table 4.3, Figure 4.4 demonstrates the predictive accuracy measured with the AUC on the simulated data with probit link for the training and the validation data. Confidence intervals are included in the figure. The box plots for the results with the other link functions are included in the Appendix (Figures A.1 to A.4). As previously described above, the ranges and medians of the AUC measures are quite similar for the two approaches for the probit link. Only a few outliers are indicated by the box plots.

Since the true distribution in the data is not known, most of the statistical models imply special model assumptions. The aim of the simulation study is to analyze different link functions in the data to compare the AUC approach with the logit model. While the AUC

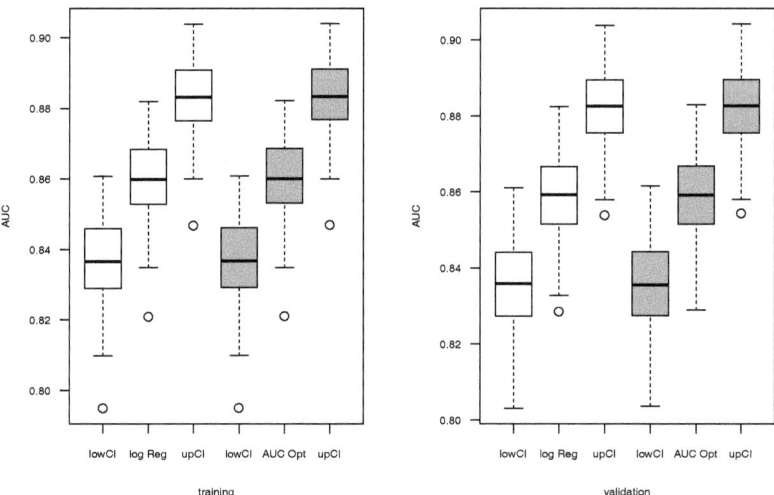

Figure 4.4: AUC comparison out of 100 simulations for the logit model (white) and the AUC approach (grey) with 95% confidence intervals on the simulated training data (left) and applied to the validation data (right) with a simulated probit link.

approach does not need an assumption for a special link function, the logistic regression model assumes the logistic function in the data. According to the simulation results, the AUC outperforms the logit model if the logistic relationship in the data is not true. In cases where the logistic model assumptions fail but still a monotone increasing function exists, the AUC approach performs better or at least equal to the logit model. The more the true link function in the data deviates from the logistic one, the greater is the improvement of the AUC approach. An important advantage of the AUC approach is that the optimality properties still are true even if there is no logistic link function.

Pepe et al. (2006) indicate similar conclusions in their paper regarding different model assumptions where they compare an AUC-based method with the logistic regression. The focus in their simulation, however, is on small data samples to demonstrate the differences in the methods.

The former simulations are based on independent normally distributed explanatory variables. In order to prove the findings and deal with realistic assumptions for real-world data, a multivariate normal distribution for the explanatory variables is analyzed to simulate correlation in the data. I follow the same simulation procedure as described previously with a bivariate response variable and three covariates, and concentrate on the AUC measure. The following covariance matrix Σ is used for generating the variables in the data

$$\Sigma = \begin{pmatrix} 1 & 0.5 & 0.5 \\ 0.5 & 1 & 0.5 \\ 0.5 & 0.5 & 1 \end{pmatrix} \tag{4.14}$$

The five link functions are used to create the training and validation data. The results for the 100 simulations for the training and validation data are shown in Table 4.4. In general, the same conclusions can be drawn as from the former simulations. Regarding the training data, the AUC approach outperforms the logit model in all cases, regardless of the underlying link function. Regarding the validation data for link2, the mean value of the AUC, for the AUC based method with 0.8353, definitely exceeds the mean value of 0.8353 for the predictive accuracy of the logit model. The AUC means equal for both methods with an AUC value of 0.9006, assuming the probit link in the data. For the logit link and the complementary log–log link, the validation outcomes are slightly better for the logit model while for link1, the AUC mean of the AUC approach lies above the predictive accuracy of the logistic regression.

	training				**validation**			
	logReg		AUC approach		logReg		AUC approach	
	mean	sd	mean	sd	mean	sd	mean	sd
logit	0.8148	0.0140	0.8150	0.0140	0.8107	0.0143	0.8106	0.0144
probit	0.9030	0.0089	0.9032	0.0089	0.9006	0.0088	0.9006	0.0087
comp log log	0.8902	0.0105	0.8903	0.0105	0.8867	0.0112	0.8866	0.0112
link1	0.9238	0.0085	0.9240	0.0085	0.9219	0.0081	0.9220	0.0081
link2	0.8345	0.0125	0.8350	0.0123	0.8353	0.0123	0.8356	0.0121

Table 4.4: AUC means and standard deviations out of 100 simulations for the logit model and the AUC approach estimated on the simulated training samples and applied to the simulated validation data for five different link functions and the covariance matrix Σ.

The simulation results approve the theoretical considerations of the previous section. The results for the independent covariates and the variables with adjusted covariance matrix are comparable. If the logistic model holds true, both methods offer optimal estimates, and according to the simulation study, the predictive accuracy of the logit model is a little higher or on the same level than for the AUC approach. If the logistic link function is not true but a monotone increasing function exists, the optimality properties for the AUC approach still hold in contrast to the logit model. The simulation results demonstrate superiority for the AUC approach in cases where the logistic model assumption fails and the link functions intensively deviate from the logistic one.

The simulation results also indicate that the logistic regression results are quite robust, even if there is no logistic link function. In cases where the link function only slightly differs from the logistic one, the simulation results demonstrate similar predictive accuracy for both approaches. Li and Duan (1989) also show robustness of the logistic regression under link violation. However, the fact that the AUC approach works without definition of a specific link function definitely qualifies the method as an alternative to the logit model.

4.3 The AUC Approach in Credit Scoring

In the following, the approach of direct AUC optimization is analyzed for the credit scoring case. Instead of using the maximum likelihood estimation, the coefficients are directly optimized with the AUC measure by employing the Wilcoxon–Mann–Whitney procedure as method of calculation for AUC and the Nelder–Mead method as the optimization algorithm. The outcomes are compared using different performance measures, and DeLong's test is used to analyze the significance of the differences of the AUC measures.

The basis for the analysis is the credit scoring model with 8 variables selected with the logit model. First, continuous explanatory variables were used. Next, I needed to remove data sets with encoded meaning. Retirees, for example, were excluded from the training sample while they are coded in the variable duration of employment and do not represent a numerical order. It is started with two variables and proceeded by adding a variable at each step.

By reason of normalization, one parameter is fixed to 1 for optimizing the others in relation to the given parameter. For the anchor parameter, the most important variable of the model was used even though the choice of the fixed variable is not important. In addition, the estimated parameters from the logit model are used as starting values for the optimization algorithm divided by the coefficient of the fixed variable. Table 4.5 shows the AUC measures for the AUC approach compared to the logistic regression results.

	train		**test**		**validation**		
	LogReg	AUC Opt	LogReg	AUC Opt	LogReg	AUC Opt	p-value
2 variables	0.7025	0.7034	0.6650	0.6660	0.6800	0.6804	0.6401
3 variables	0.7082	0.7098	0.6753	0.6751	0.6951	0.6943	0.4466
4 variables	0.7104	0.7141	0.6793	0.6801	0.6939	0.6969	0.0335
5 variables	0.7140	0.7174	0.6835	0.6830	0.6985	0.7002	0.3171
6 variables	0.7150	0.7183	0.6841	0.6838	0.6993	0.7007	0.3020
7 variables	0.7344	0.7357	0.7051	0.7048	0.7125	0.7141	0.0248

Table 4.5: AUC optimization by calculation with the Wilcoxon–Mann–Whitney and Nelder–Mead algorithms for the 50% training sample, applied to the test and validation samples with continuous explanatory variables. The p-values result from DeLong's test.

For the training sample, logit model outcomes can be outperformed by directly optimizing the AUC even though the improvement is slight. While the optimization results and the logit model outcomes for the test sample are at the same level, the validation results show a little improvement for the AUC approach. Most of the results on the validation data are , however, not significant in terms of DeLong's test.

Table 4.6 shows the results for different performance measures. In most cases, the AUC approach outperforms the logit model even if the differences are small. The Brier score indicates lower predictive performance on the 50% training sample, but denotes higher performance on the test and validation samples. These empirical results are similar to the outcomes of the simulation study (Table 4.3).

		train		test		validation	
		LogReg	AUC Opt	LogReg	AUC Opt	LogReg	AUC Opt
2 variables	AUC	0.7025	0.7034$^{(1)}$	0.6650	0.6660$^{(1)}$	0.6800	0.6804$^{(1)}$
	H	0.1515	0.1519$^{(1)}$	0.0828	0.0856$^{(1)}$	0.0953	0.0965$^{(1)}$
	KS	0.3091	0.3110$^{(1)}$	0.2531	0.2544$^{(1)}$	0.2799	0.2770$^{(1)}$
	MER	0.3452	0.3442$^{(1)}$	0.0275	0.0275$^{(1)}$	0.0272	0.0271$^{(1)}$
	Brier	0.2239$^{(2)}$	0.2658	0.2495	0.1011$^{(1)}$	0.2540	0.1017$^{(1)}$
3 variables	AUC	0.7082	0.7098$^{(1)}$	0.6753$^{(2)}$	0.6751	0.6951	0.6943$^{(1)}$
	H	0.1613	0.1623$^{(1)}$	0.0865	0.0891$^{(1)}$	0.1086	0.1097$^{(1)}$
	KS	0.3176	0.3185$^{(1)}$	0.2720$^{(2)}$	0.2662	0.3050	0.3051$^{(1)}$
	MER	0.3398	0.3376$^{(1)}$	0.0275	0.0275$^{(1)}$	0.0272	0.0271$^{(1)}$
	Brier	0.2197$^{(2)}$	0.2796	0.2501	0.0854$^{(1)}$	0.2579	0.0872$^{(1)}$
4 variables	AUC	0.7104	0.7141$^{(1)}$	0.6793	0.6801$^{(1)}$	0.6939	0.6969$^{(1)}$
	H	0.1629	0.1654$^{(1)}$	0.0927	0.0974$^{(1)}$	0.1049	0.1124$^{(1)}$
	KS	0.3161	0.3292$^{(1)}$	0.2666	0.2716$^{(1)}$	0.2896	0.3062$^{(1)}$
	MER	0.3385	0.3350$^{(1)}$	0.0275	0.0275$^{(1)}$	0.0272	0.0272$^{(1)}$
	Brier	0.2181$^{(2)}$	0.2911	0.2457	0.0754$^{(1)}$	0.2534	0.0766$^{(1)}$
5 variables	AUC	0.7140	0.7174$^{(1)}$	0.6835$^{(1)}$	0.6830	0.6985	0.7002$^{(1)}$
	H	0.1688	0.1716$^{(1)}$	0.0954	0.1009$^{(1)}$	0.1107	0.1175$^{(1)}$
	KS	0.3300	0.3374$^{(1)}$	0.2746$^{(1)}$	0.2744	0.2952	0.3084$^{(1)}$
	MER	0.3331	0.3315$^{(1)}$	0.0275	0.0275$^{(1)}$	0.0272	0.0272$^{(1)}$
	Brier	0.2169$^{(2)}$	0.2945	0.2461	0.0720$^{(1)}$	0.2549	0.0734$^{(1)}$
6 variables	AUC	0.7150	0.7183$^{(1)}$	0.6841$^{(1)}$	0.6838	0.6993	0.7007$^{(1)}$
	H	0.1701	0.1733$^{(1)}$	0.0970	0.1022$^{(1)}$	0.1122	0.1186$^{(1)}$
	KS	0.3263	0.3370$^{(1)}$	0.2734	0.2735$^{(1)}$	0.2966	0.3091$^{(1)}$
	MER	0.3347	0.3325$^{(1)}$	0.0275	0.0275$^{(1)}$	0.0272	0.0272$^{(1)}$
	Brier	0.2165$^{(2)}$	0.3050	0.2455	0.0640$^{(1)}$	0.2544	0.0653$^{(1)}$
7 variables	AUC	0.7344	0.7357$^{(1)}$	0.7051$^{(1)}$	0.7048	0.7125	0.7141$^{(1)}$
	H	0.1986	0.2009$^{(1)}$	0.1211	0.1219$^{(1)}$	0.1276	0.1302$^{(1)}$
	KS	0.3662	0.3686$^{(1)}$	0.3065$^{(1)}$	0.3053	0.3142	0.3177$^{(1)}$
	MER	0.3140$^{(2)}$	0.3144	0.0275	0.0274$^{(1)}$	0.0272	0.0272$^{(1)}$
	Brier	0.2099$^{(2)}$	0.3857	0.2380	0.0332$^{(1)}$	0.2494	0.0337$^{(1)}$

Table 4.6: Different measures for the AUC optimization using the Wilcoxon–Mann–Whitney and Nelder–Mead algorithms for the 50% training sample, applied to the test and validation samples with continuous explanatory variables. (1) denotes cases where the AUC approach outperforms or is equivalent. (2) denotes cases where the logit model presents better results compared to the AUC approach.

The estimation with 7 variables shows validation results with an AUC value of 0.7125 for the logit model compared to 0.7141 for the AUC approach (Table 4.6). The H measure denotes values of 0.1276 and 0.1302 while the KS indicates better performance for the AUC approach with a value of 0.3177 (compared to 0.3142 for the logit model).

The analysis of the AUC approach was started with two parameters, the coefficient of the duration of employment was fixed to 1, and an income variable together with a variable of indebtedness were optimized. While the β-coefficients for the logit model $\hat{\beta}_{ML} = (1.270784, -0.005108, -0.000408, -0.000518)$ result in an AUC value of 70.25%, the estimated β-coefficients of the AUC approach $\hat{\beta}_{AUC} = (-1, -0.042633, -0.074214)$ yield a performance measure of 70.34%. Note that the calculation method requires no constant, and the first parameter is fixed. The best improvements appear by 5 to 7 optimized variables. Because of computational reasons, it is concentrated on the training sample with 50% default rate and applied it to the test and validation samples.

The visualization of the optimization can be seen in Figure 4.5, which contains two β-coefficients for the 50% sample. The plot shows β_1 and β_2 for the x- and y-axis respectively, while the z-axis shows values of the belonging $AUC(\beta)$.

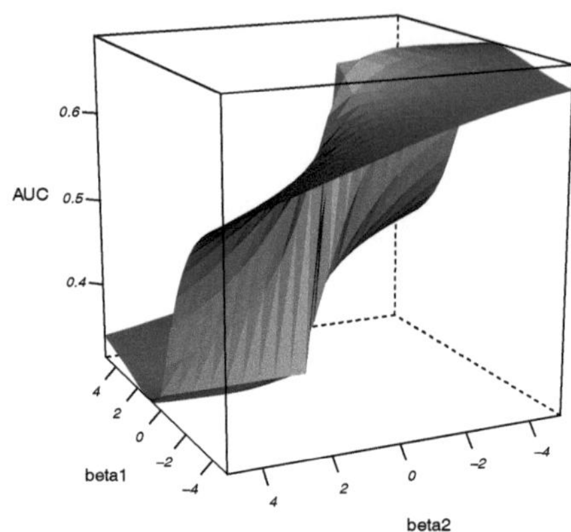

Figure 4.5: Optimization of the β-coefficients with the Wilcoxon statistic and the Nelder–Mead algorithm. Visualization of the coefficients from two explanatory variables (x-axis and y-axis) and the $AUC(\beta)$ values (z-axis).

In addition, the AUC approach is also tested for categorical variables. The procedure is the same as described above. The 50% training sample is used, one parameter is fixed, and it is applied to the test and validation samples. The results are shown in Table 4.7.

	train			test			validation		
	LogReg	**AUCOpt**	MaxDiv	LogReg	**AUCOpt**	MaxDiv	LogReg	**AUCOpt**	MaxDiv
AUC	0.7089	0.7099	0.6892	0.6661	0.6669	0.6506	0.6770	0.6787	0.6659
p-value							0.0104		0.0000
H measure	0.1585	0.1585	0.1375	0.0860	0.0869	0.0682	0.0934	0.0954	0.0790
KS	0.3200	0.31200	0.2988	0.2578	0.2584	0.2481	0.2686	0.2689	0.2644
MER	0.3400	0.3400	0.3506	0.0263	0.0263	0.0263	0.0260	0.0260	0.0260
Divergence	0.6116	0.6328	0.7071	0.3634	0.4028	0.3158	0.4195	0.4434	0.4445

Table 4.7: AUC optimization using the Wilcoxon–Mann–Whitney and Nelder–Mead algorithms for the 50% training sample, applied to the test and validation samples with 3 classified explanatory variables in comparison to the logit model and the maximization of the divergence. The *p*-values result from DeLong's test and investigate the AUC differences compared to the AUC approach.

An important aspect of using categorical variables is the creation of dummy variables for each particular class. Except for one reference class, one 1/0-variable is created for each class of a variable. This is similar to the procedure for logistic regression. The test is limited to three variables with five, four and four categories and, therefore, the estimation of 10 parameters. In addition, the AUC approach is compared to the approach of Thomas (2009) for maximizing divergence. Table 4.7 shows the comparisons of the three methods.

The coefficient values of the logistic regression are used as starting values for the AUC optimization as well as for the divergence approach. For the three categorical variables, 10 parameters are estimated. Regarding the AUC values in Table 4.7, while the AUC approach is slightly better than the logit model, it definitely outperforms the divergence approach. The p-values result from DeLong's test and denote the test in comparison to the AUC approach. According to this test, both differences are significant on the validation data. Considering divergence as a performance measure, the divergence approach outperforms for both the training sample and the validation sample, but shows lower performance on the test data. Concerning the H measure, the KS and the MER, the AUC approach shows superior results compared to the divergence approach, mostly providing outperforming outcomes compared to the logit model.

4.4 Discussion of the AUC Approach in Credit Scoring

Since the AUC is a prominent performance measure in the credit scoring area, an approach is introduced to optimize this measure directly. The relationship to the well-known Wilcoxon statistic is used, as well as the Nelder–Mead algorithm as optimization algorithm. The theoretical considerations are pointed out and also the relationship to the maximum rank correlation estimator. Besides this, a simulation study is conducted to examine the properties showing slight superiority for the AUC approach if the model assumption fails for the logistic model but a monotone increasing function denotes the data.

The proposed AUC approach seems to be a reasonable procedure for estimating the parameters indicating slight improvements for the credit scoring case compared to the classical logit model. A main advantage of the proposed method is that it does not imply a special link function. Since the logit model assumption fails for the logistic regression, there are no optimality properties, while the AUC approach still has the optimal ROC curve among all other linear combinations. Therefore, the theoretical considerations hold even if the link function is not logistic but a monotone increasing function.

However, the method is actually computationally intensive and constrains the analysis to a limited number of data sets. Since there are large data sets in credit scoring, extending the approach for application on big data is a point for further research.

Note that the considerations of the AUC approach do not hold for non linear functions or complex interactions. For covering these topics, random forests and boosting algorithms are presented in this thesis.

Chapter 5

Generalized Additive Model

5.1 A Short Overview of Generalized Additive Model

For building scorecards in credit scoring, the logistic regression model is the most widely used method. This approach presented in Chapter 3 belongs to the group of generalized linear models (GLMs). Since classical linear models assume a continuous, and at least approximately normal distributed response, GLMs are extended to non-normal response variables. For binary outcomes, the logit model is a prominent example of generalized linear models to analyze the probability $\pi_i = P(Y_i = 1 \mid x_i) = G(x_i'\beta)$ for the binary response Y_i (default and non-default). The effect of the covariates to the response is covered with a linear predictor. The relationship between the mean of the binary response and the predictor variables is defined with the following logit link function:

$$\log\left(\frac{\pi_i}{1-\pi_i}\right) = \beta_0 + \beta_1 x_1 + \ldots + \beta_p x_p \tag{5.1}$$

An extension to the GLMs are the so-called generalized additive models (GAMs) for capturing nonlinear regression effects (Hastie and Tibshirani, 1990). The relationship between the response and the explanatory variables can also be defined through an arbitrary link function like, for instance, logit or probit. Therefore, the additive logistic regression model follows the form

$$\log\left(\frac{\pi_i}{1-\pi_i}\right) = \beta_0 + f_1(x_1) + \ldots + f_p(x_p) \tag{5.2}$$

where the f_j's denote unspecified smooth functions. The additivity of the model allows the interpretability in a similar way as before and the nonparametric form of the functions f_j adds more flexibility to the model (Hastie et al., 2009). By using an algorithm based on a scatterplot smoother, the estimation of the functions f_j is flexible. An advantage is that linear and nonlinear effects can be mixed in the model what is relevant if categorical covariates are included in the regression. A semiparametric model results by modelling linear predictors in combination with qualitative variables and nonparametrically predictors.

The estimation in generalized additive models replaces the weighted linear regression by a weighted backfitting algorithm (Hastie et al., 2009). This is the so-called *local scoring algorithm.* It represents an iterative algorithm by iteratively smoothing partial residuals. For the logistic regression case, the Newton-Raphson algorithm used for maximizing log-likelihoods for generalized linear models is altered as an iteratively reweighted least squares (IRLS) algorithm. In the linear case, a weighted linear regression of a working outcome variable on the covariates is fitted (Hastie et al., 2009). Since in each iteration new parameter estimates are received, it results in new working responses, computed weights, and further iteration of the process. As mentioned above, the weighted linear regression is replaced by using a weighted backfitting algorithm for estimating smoothing components. Details for the backfitting algorithm for additive models and the local scoring algorithm for the additive logistic regression model can be found in Hastie and Tibshirani (1990) and Hastie et al. (2009).

For the presented credit scoring case, GAMs are used within the presented framework of backfitting with local scoring. Another approach is considered by Wood (2006) using penalized regression splines estimated by penalized regression methods.

5.2 Generalized Additive Model for Credit Scoring

For the binary problem in credit scoring, an additive logistic regression is applied to the consumer credit data. Since the emphasis of this thesis lies in the new statistical methods, the generalized additive model is analyzed based on the logistic regression presented in Section 3.2. Without variable selection, it is tried to improve the logit model by analyzing smoothing effects for the covariates.

The generalized additive model was fit, keeping one explanatory variable as factor, while using smoothing splines for the other seven covariates. Hastie et al. (2009) propose a nominal four degrees of freedom as a convenient way for the specification of the amount of smoothing of the predictor variables. This implies that the smoothing-spline parameter λ_j is chosen so that the smoothing spline operator matrix $S_j(\lambda)$ follows $\text{trace}[S_j(\lambda_j)] - 1 = 4$.

The additive logistic regression model is estimated on each of the three training samples and applied the results to the test and validation sample. Note that observations with encoded meaning are eliminated in the data sets since continuous covariates are used. On all three training data sets, the smoothing predictors are significant in the additive model. Table 5.1 shows the predictive performance of the estimated additive models on the different data samples. Furthermore, the AUC values for the logit models with continuous covariates are displayed to compare the prediction accuracy.

As expected, the highest predictive accuracy is realized for the training samples. For instance, on the 20% training sample, the AUC measure denotes a value of 0.7606. On the test and validation sample, the models estimated on different training samples yield a higher predictive accuracy for the additive logistic regression model than for the linear logit model. The outcomes on the validation sample for the models estimated on the 50% training sample are 0.7263 for the additive model compared to 0.7125 for linear logit model.

	AUC	lowCI95%	upCI95%	LogReg	p-value	
trainORG	0.7561	0.7466	0.7656	0.7371		trainORG
	0.7210	0.7112	0.7308	0.7045		test
	0.7250	0.7115	0.7386	0.7118	0.1068	validation
train020	0.7606	0.7491	0.7721	0.7405		train020
	0.7205	0.7107	0.7303	0.7054		test
	0.7250	0.7115	0.7385	0.7129	0.1365	validation
train050	0.7562	0.7393	0.7731	0.7344		train050
	0.7207	0.7109	0.7305	0.7051		test
	0.7263	0.7128	0.7399	0.7125	0.0879	validation

Table 5.1: Prediction accuracy of the additive logistic regression models estimated on the training samples and applied to the test and validation sample measured with the AUC and continuous covariates. The p-values result from DeLong's test.

Similar conclusions can be drawn from the validation results for the models estimated on the 20% training sample and the original training sample. The AUC values of 0.7250 (train020) and 0.7250 (trainORG) for the additive logistic regression are higher than the predictive accuracy for the logit model with 0.7129 (train020) and 0.7118 (trainORG). The results for the logit models are realized by using continuous variables as covariates. According to DeLong's test, however, the differences between the AUC measures on the validation data are not significant.

Table 5.2 shows different performance measures for the validation results. The generalized additive models are estimated on the training sample and applied to the validation data. All measures indicate the same conclusions as for the AUC results. The H measures and KS values are higher for the generalized additive model than for the logit model. The minimum error rates are on the same level, and the Brier scores are lower for the GAM except the results trained on the original training sample.

		AUC	Gini	H measure	KS	MER	Brier score
trainORG	GAM	0.7250	0.4501	0.1404	0.3338	0.0272	0.0270
	LogReg	0.7118	0.4237	0.1283	0.3101	0.0272	0.0270
train020	GAM	0.7250	0.4500	0.1403	0.3313	0.0271	0.0703
	LogReg	0.7129	0.4257	0.1286	0.3143	0.0272	0.0719
train050	GAM	0.7263	0.4527	0.1414	0.3314	0.0272	0.2353
	LogReg	0.7125	0.4249	0.1276	0.3142	0.0272	0.2494

Table 5.2: Further measures for the GAM results compared to the logistic regression results with continuous variables for the validation data.

For the validation results trained on the 20% training sample, the H measure denotes 0.1403 for the generalized additive model compared to 0.1286 for the logit model. The KS denotes 0.3313 and 0.3143 for the GAM and the logit model, respectively. The Brier score for the validation data trained on the 20% training data is lower and, therefore, better for the GAM (0.0703 to 0.0719 for the logit model).

The additive logistic regression model improves the predictive performance of the linear logit model. The flexibility of the model and the inclusion of smoothing functions results in higher AUC measures of the additive model, which can be illustrated for the corresponding ROC curves. Figure 5.1 shows the ROC graphs for the logit model and the additive model on the validation data. As shown in the figure, the dashed ROC curve of the logit model with continuous variables lies beneath the curve of the additive logistic regression.

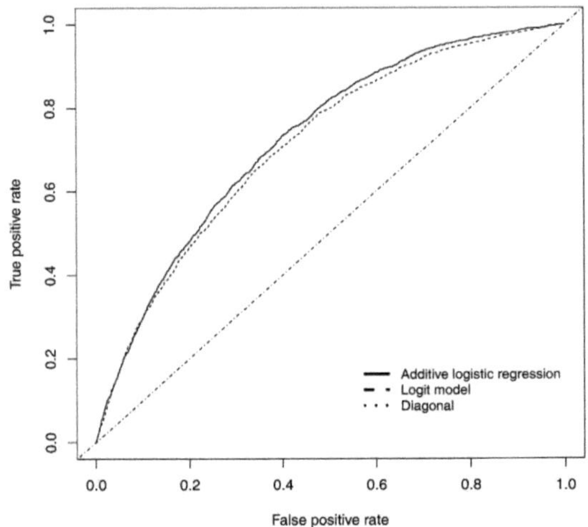

Figure 5.1: ROC graphs for the additive logistic regression (AUC 0.7250) and the logit model with continuous covariates (AUC 0.7118) estimated on the original training sample applied to the validation data.

The flexibility of the GAMs and the inclusion of nonlinearities in the additive model has an additional effect on the predictive performance for the credit scoring data. Figure 5.2 shows the significant predictors in the additive logistic model estimated on the 50% training sample. Seven variables display smoothing effects while one explanatory variable is included as factor. Due to confidentiality reasons, the distributions of the variables can not be interpreted in detail. The estimated functions shown in Figure 5.2 clearly indicate nonlinearities. Note that the rug plot along the bottom of each illustration indicates the observed values of the corresponding predictor. Variable 7, for example, shows nonlinearity for only a few observations at the top end of the predictor, while variable 25 seems to be rather linear distributed for the area where most of the observations are represented.

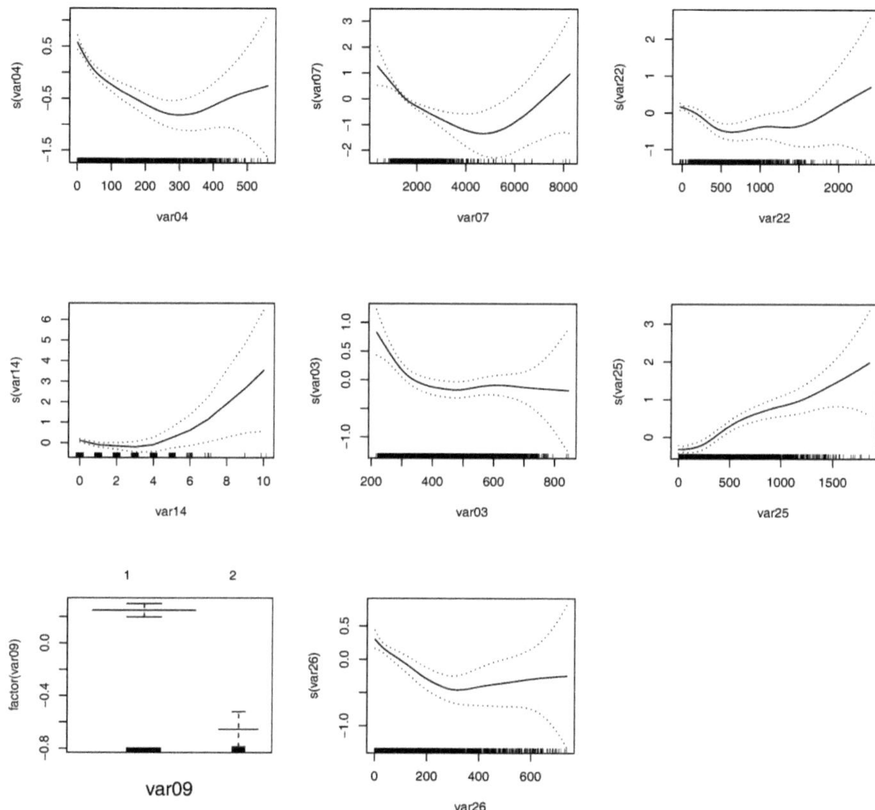

Figure 5.2: Estimated functions of seven continuous predictor variables and one factor variable of the additive logistic regression model estimated on the original training sample. The observations of the corresponding predictors are displayed on the bottom of the x-axis. The dashed lines present the upper and lower pointwise twice-standard-error curves. Nonlinearities are indicated for most of the predictors.

The additive logistic regression model also outperforms the classical logit model with classified explanatory variables. The predictive accuracy for the validation data yields an AUC value of 0.7224 for the logit model with categorized explanatory variables estimated on the 50% training data. The corresponding AUC measure for the additive model denotes a higher predictive performance with 0.7263. This is the same for the validation results for the model estimation on the 20% training data and the original training sample (GAM 0.7250 to LogReg 0.7237; GAM 0.7250 to LogReg 0.7219).

Some of the performance measures, however, indicate poorer results for the generalized additive model compared to the logit model with coarse classification. The H measure for the generalized additive model trained on the 20% training sample denotes 0.1403 for the validation data compared to 0.1428 for the logit model with classified variables. The other performance measures, including KS and Brier score, indicate better performance for the generalized additive model. The comparison of the GAM results to the logit model results with classified covariates is drawn in Figure A.1 in the Appendix.

5.3 Discussion of a Generalized Additive Model in Credit Scoring

The evaluation of the additive logistic regression model for the presented credit scoring case shows outperforming outcomes compared to the classical logit model. The flexibility of the model and the inclusion of nonlinearities in the method add performance to the predictive accuracy measured by the AUC and different performance measures.

A main advantage of the additive logistic regression model is that it provides information about the relationship between the response and each predictor variable not revealed by the use of standard modeling techniques. The additive form of the approach still offers familiar tools for modeling and inference comparable to the linear models. Compared to other nonparametric algorithms, the generalized additive model allows analysis and interpretation of the influence of each single predictor variable.

The presented results imply promising improvements for the predictive performance. By including all potential variables and more smoothing effects, the performance can probably be improved. A procedure for variable selection would then be needed. However, the additive model and the consideration of smoothing splines in the model adds more complexity to the model. Even if the additivity of the model and the underlying backfitting algorithm comes along with good interpretability compared to other nonparametric approaches, the extension of the linear model represents a more complicated model than the linear version. Simple explanation is still a very important aspect in credit scoring practice.

GAM procedures are implemented in many standard statistical software packages. Therefore, it is easy for financial institutions to implement and analyze GAMs for the credit scoring case. The algorithm can handle great data sets with a huge amount of observations about what is also relevant for credit scoring. The additive logistic model represents a good alternative for analyzing the predictive accuracy in credit scoring. And even if it does not replace the classical scorecard development with the logistic regression, it represents a useful instrument for detecting nonlinearities and relationships between the default probability and the explanatory variables. These investigations could then lead to variable transformations, which can be finally included in the classical logit model.

Chapter 6

Recursive Partitioning

6.1 Classification and Regression Trees

In recent years, recursive partitioning algorithms have become very popular in scientific fields like biostatistics and medicine. The aim is to analyze these newer methods in the credit scoring context in order to maximize the predictive accuracy with regard to the AUC measure.

The Classification and regression tree algorithm (CART) of Breiman et al. (1984) and the C4.5 algorithm of Quinlan (1986, 1993) are the most well-known splitting methods. For the general descriptions, it is concentrated on the CART method. The recursive partitioning algorithm presents a nonparametric approach without an assumption of a special distribution (Strobl et al., 2009). In contrast to classical regression models, the partition in classification trees allows nonlinear and nonmonotone association rules without advance specification. The method is binary and recursive; i.e., every root node is divided into two groups, and the arising end nodes are divided again. Figure 6.1 represents an exemplary illustration for credit scoring data with a binary response (default and non-default) and predictor variables x_i, x_j and x_k.

First, the whole data set is split into two regions while the variable and the best split-point are chosen according to the best fit. In the example, input variable x_i is split at $x_i < c_i$ and $x_i \geq c_i$. One or both arising regions are divided further into two more regions. This procedure is repeated until a stopping criterion is reached. While observations for $x_i < c_i$ are not divided further, observations which follow the rule $x_i \geq c_i$ are split with the predictor variable x_j at the split-point c_j. Observations satisfying the condition $x_j < c_j$ are divided following the variable x_k for the split $x_k < c_k$ and $x_k \geq c_k$. The main idea for finding a split is to choose the binary partition in the way that the arising child nodes are purer and more homogeneous than the parent nodes. In the example, the terminal nodes are labeled with good credit and bad credit in order to show the majority of either non-defaults or defaults in the node. Each of these nodes contains a number of non-defaults and defaults, respectively.

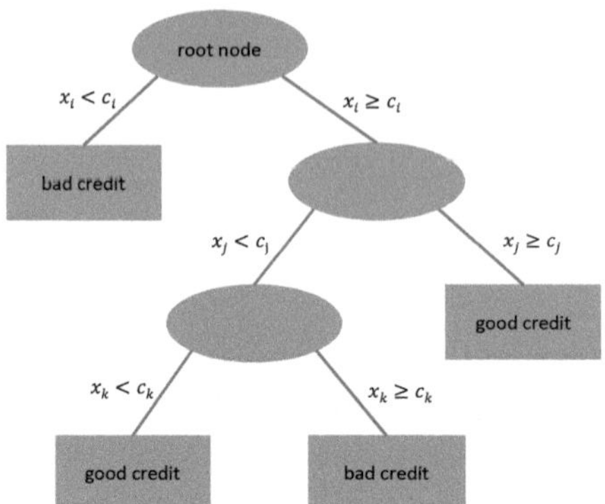

Figure 6.1: Example of a classification tree with binary splits.

Another exemplary illustration for a binary classification tree with four splits is shown in Table 6.2 with the input variables var03 and var20 from the credit scoring data. Observations with values for the variable var03, which exceed or equal 326, are split into the left child node. They are divided further with the split point 448. The others are split in the right node representing a terminal node. Since there are only two potential variables for splitting the tree, both variables are used more often. The tree is exemplarily grown for 1000 observations and the number of defaults and non-defaults in the nodes are also displayed. The root node contains 500 defaults and 500 non-defaults. Split into two child nodes, the left one includes 431 non-defaults and 330 defaults. The right branch indicates 69 non-defaults and 170 defaults for the terminal node.

Since the full partition of the feature space can be described with a single tree, the interpretability denotes an advantage of recursive binary trees (Hastie et al., 2009). The graphical illustration for a lot of input variables and complex trees is difficult, but the basic principle remains the same.

6.1.1 CART Algorithm

The aim of a classification algorithm is to decide about the splitting variables and split points. The following descriptions for the CART-algorithm are based on the work of Breiman et al. (1984) and the explanations of Hastie et al. (2009).

Depending on the outcome variable, different splitting rules are proposed. In the regression case with continuous response, the splitting criterion is determined by minimization of the residual sum of squares. For a binary response, the Gini index is mainly used as a measure of node impurity. In the two class case, the Gini index is defined as

$$2p(1-p) \tag{6.1}$$

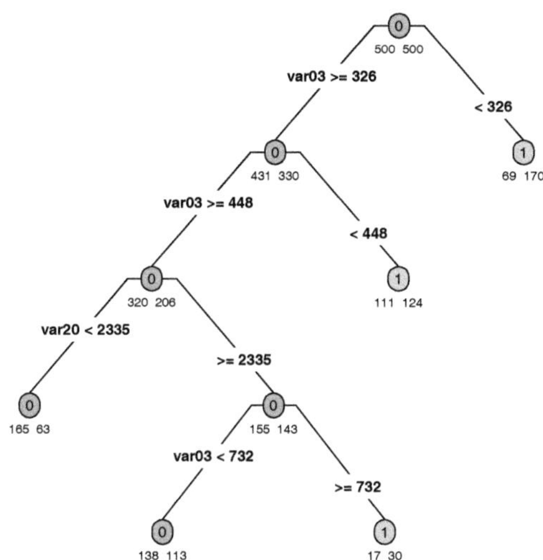

Figure 6.2: Illustration of a classification tree on credit scoring data with two explanatory variables. 0 and 1 denote non-defaults and defaults, respectively. The values under each node represent the non-defaults and defaults in a node. The number of observations in the example is 1000 with a 50% default rate.

with p as proportion in the second class. For the credit scoring case, this would imply the proportion of defaults in a node. The Gini index specifies the probability that an object in a node is falsely classified. The impurity measure receives values between 0 and 0.5 where 0 represents nodes with only one response class and therefore totally pure nodes. A value of $1/2$ denotes maximal impurity in a node since both response classes are equally represented in the node. The chosen splitting rule follows the approach of impurity reduction in the way that the daughter nodes are purer than the parent nodes. The best split of all possibilities is found according to the impurity reduction. For measuring the impurity reduction, the difference between the impurity in the parent node and the average impurity in the two daughter nodes is used (Strobl et al., 2009).

Other measures of node impurity are the misclassification error and the cross-entropy, which are defined for the two class case as $1-\max(p, 1-p)$ and $-p\log p-(1-p)\log(1-p)$, respectively. Compared to the misclassification error, the Gini index and the cross-entropy are differentiable and more sensitive against changes in the node probabilities. Since both measures should be prefered in contrast to the misclassification error, it is concentrated on the Gini index for the credit scoring analysis.

The process of recursively splitting the variables and finding the optimal cutpoint is repeated until a special stopping criterion is reached. The question arises when to stop growing the tree. Different criteria are proposed; a given threshold for the impurity measure that is not exceeded by any split, total purity for terminal nodes, and a minimum number of observations in a node for splitting it further. The first criterion is simple, yet critical, since it is difficult to define a threshold for all splits, and since a poor split can have a good split afterwards. The prefered strategy is to grow large trees and use cost-complexity pruning to find the optimal tree size and to prevent overfitting (Hastie et al., 2009). A subtree $T \subset T_0$ is received by pruning the tree T_0. Since m denotes the index for terminal nodes, N_m the number of observations and $|T|$ the number of terminal nodes in T, the definition for the cost complexity criterion follows:

$$C_\alpha(T) = \sum_{m=1}^{|T|} N_m Q_m(T) + \alpha|T| \tag{6.2}$$

For the cost complexity pruning in classification, the misclassification error is typically used as impurity measure $Q_m(T)$. The aim is to find the subtree $T_\alpha \subseteq T_0$ which minimizes $C_\alpha(T)$. Since the emphasis for the credit scoring evaluation is to find the best predictive accuray regarding the AUC measure, the AUC is used in the analysis to find the optimal tree size.

A value of $\alpha = 0$ grows a full tree, while large values of α lead to smaller trees. The tuning parameter α is therefore relevant for the tradeoff between the tree size and the fitting to the data. It can be shown that there is a unique smallest subtree T_α which minimizes $C_\alpha(T)$ for each α. Details can be found in Breiman et al. (1984) and Hastie et al. (2009). The estimation of α is yielded with cross-validation.

Another tuning parameter for the classification algorithm represents the loss matrix. In many classification problems, the misclassification of an observation in one class is more serious than in the other. In credit scoring, for instance, it is worse for a bank to classify a customer as a non-default when the credit amount can not be paid back, than vice versa. The loss parameter denotes a factor to weight the misclassification in one class more serious than in the other. This is defined with a loss matrix L where $L_{kk'}$ represents the loss for incorrectly classifying a class k observation as class k'. In the two class case, the observations in class k are weighted with $L_{kk'}$ (Hastie et al., 2009). To alter the prior probability on the classes is the effect of this observation weighting.

For prediction, the final response class is either calculated by the most frequent response class or by the relative frequencies of each class in the terminal nodes (Strobl et al., 2009). Several works, however, indicate that classification trees provide poor probability estimates. Provost and Domingos (2003), for instance, propose the common Laplace correction for improving the accuracy and the growing of larger trees in order to achieve better probability estimates. Ferri et al. (2003) propose different smoothing approaches and split criteria including the AUC. The latter results show no improvements by changing the split criterion. An AUC based classification algorithm is shown in Zhang and Su (2006) comparing different classification tree algorithms with each other.

For the credit scoring case, I investigate the following two smoothing techniques regarding the best AUC measure (Ferri et al., 2003):

$$\text{Laplace smoothing } p_i = \frac{n_{nd} + 1}{n + c} \tag{6.3}$$

$$\text{m-estimate smoothing } p_i = \frac{n_{nd} + m \cdot p}{n + m} \tag{6.4}$$

where n_{nd} denotes the number of non-defaults in a terminal node, n the observations in a terminal node and c the number of classes. In the presented case, it is a two class problem with $c = 2$. Therefore, the Laplace smoothing is a special case of the m-estimate smoothing with $m = 2$ and $p = 0.5$. In the m-estimate smoothing, a number of m additional observations is added, while p denotes the expected probability of a non-default.

Breiman et al. (1984) already noted that variables with many categories and continuous variables are favored during the partition process. This is the same in other classification methods like C4.5 (Quinlan, 1993) which can lead to overfitting. As a result, several algorithms with unbiased variable selection have been proposed (cf. Dobra and Gehrke (2001), Strobl et al. (2007), Hothorn et al. (2006)).

A main disadvantage of binary classification trees is the instability to small changes in the learning data. The structure of the tree algorithm determines this effect. A chosen variable and a chosen split point for splitting a node determine the partition in the next nodes. This indicates that an error in a top node is brought forward to the splits in the further nodes. But the exact splitting rule depends mainly on the data that is used to train the algorithm. Therefore, a small change in the learning data can lead to a different splitting variable or a different cutpoint, which in turn changes the whole tree structure. To overcome this instability, an alternative is to average a variety of trees leading to ensemble methods like random forests. This topic is covered in Section 6.3.

6.1.2 Conditional Inference Trees

Former tree classification algorithms, like CART or C4.5, denote two major weaknesses; overfitting (which can be solved with pruning), and the favor of variables with many potential cut-points and missing values. To overcome these problems, Hothorn et al. (2006) propose conditional inference trees with unbiased variable selection.

The approach of conditional inference trees consists of two main steps. The first step is concerned with the relation between the covariates and the response. The global null hypothesis of independence is tested between any of the input variables and the response. The fact that the hypothesis cannot be rejected at a specified nominal level α is used as stopping criterion. Otherwise the variable with the strongest relation to the response is chosen. This association is measured with the p-value of the test for the partial null hypothesis of a single covariate and the outcome variable. In the second step, the cutpoint for the selected variable from step 1 is chosen. The first and the second step are then recursively repeated.

Since the approach introduces p-values for the variable selection and for stopping the algorithm, pruning is not required in contrast to the early classification algorithms. Additionally, the separation of the variable selection and the determination of the splitting points into two steps prevent favoring variables with many potential splitting possibilities.

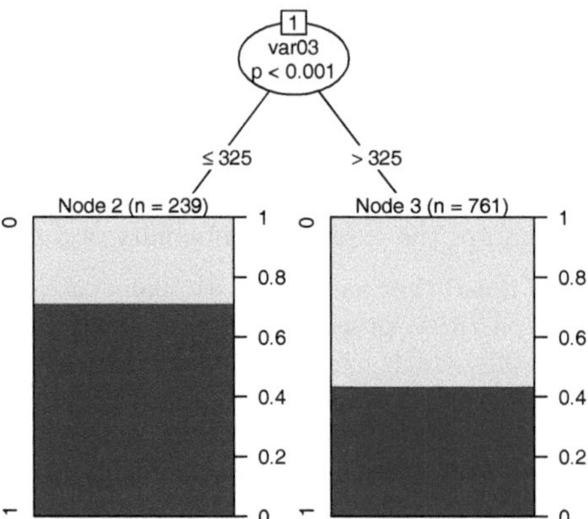

Figure 6.3: Example of a conditional inference tree on credit scoring data with two terminal nodes. 0 and 1 denote non-defaults and defaults, respectively.

Figure 6.3 shows a conditional inference tree for the credit scoring example with 1000 observations. The tree contains two terminal nodes that are produced by splitting the root node into two branches with values var03 $\leq$ 325 and var03 $>$ 325. The dark and light shaded areas in the terminal nodes display the relative frequencies of the defaults and non-defaults, respectively. In the left node, 70% of the 239 observations are defaults. The p-values of the permutation tests for the splits are also given in the graphical illustration.

A short overview of the generic algorithm in the conditional inference framework proposed in Hothorn et al. (2006) is given below. Detailed explanations can be found in the mentioned paper. The work of Hothorn et al. (2006) is based on the permutation tests and the permutation test framework developed by Strasser and Weber (1999).

As already stated above, the binary split procedure contains two steps. First, the relation between the response Y and any of the covariates $X_j, j = 1, \ldots, m$ is considered where the m-dimensional vector $\mathrm{X} = (X_1, \ldots, X_m)$ results from a sample space $\mathcal{X} = \mathcal{X}_1 \times \cdots \times \mathcal{X}_m$. The association is investigated with the m partial hypotheses H_0^j: $D(Y \mid X_j) = D(Y)$. The stop criterion is based on multiplicity adjusted p-values (e.g. Bonferroni adjustment). Since the criterion is maximized, the 1 - p-value is considered. In order to implement a split, the p-value must therefore be undercut. The form of the linear statistics for the relation between the response Y and the covariates $X_j, j = 1, \ldots, m$ follows:

$$T_j(\mathcal{L}_n, \mathbf{w}) = \text{vec}\left(\sum_{i=1}^{n} w_i g_j(X_{ji}) h(Y_i, (Y_1, \dots, Y_n))^\top\right) \in \mathbb{R}^{p_j q} \tag{6.5}$$

where $g_j \colon \mathcal{X}_j \to \mathbb{R}^{p_j}$ represents a non-random transformation of the covariate X_j. $\mathcal{L}_n = \{(Y_i, X_{1i}, \dots, X_{mi}); i = 1, \dots, n\}$ denotes the learning sample and $\mathbf{w} = (w_1, \dots, w_n)$ are case weights, which represent the correspondence of the observations to the nodes. Each tree node is indicated by a vector of case weights. The weights have non-zero positive integer values when the related observations are elements of the node and they are zero otherwise. In a permutation symmetric way, the influence function $h \colon \mathcal{Y} \times \mathcal{Y}^n \to \mathbb{R}^q$ depends on the responses $(Y_1, \dots, Y_n)$. By column-wise combination, a $p_j \times q$ matrix is converted into a $p_j q$ column vector.

The distribution of $T_j(\mathcal{L}_n, \mathbf{w})$ is depending on the joint distribution of Y and X_j. Under the null hypothesis H_0^j, by fixing the covariates and by conditioning on all possible permutations $\sigma \in S(\mathcal{L}_n, \mathbf{w})$ of the responses, the conditional expectation $\mu_j \in \mathbb{R}^{p_j q}$ and the covariance $\Sigma_j \in \mathbb{R}^{p_j q \times p_j q}$ arise as developed by Strasser and Weber (1999). With the conditional expectation μ and the covariance Σ two standardized test statistics are proposed of the form

$$c_{max}(\mathbf{t}, \mu, \Sigma) = \max_{k=1,\dots,pq} \left| \frac{(\mathbf{t} - \mu)_k}{\sqrt{(\Sigma)_{kk}}} \right| \tag{6.6}$$

or

$$c_{quad}(\mathbf{t}, \mu, \Sigma) = (\mathbf{t} - \mu)\Sigma^+(\mathbf{t} - \mu)^\top \tag{6.7}$$

mapping an observed multivariate linear statistic $\mathbf{t} \in \mathbb{R}^{pq}$ into the real line. Since the test statistic c_{max} is asymptotically normal distributed, the quadratic form c_{quad} follows an asymptotic χ^2 distribution with degrees of freedom given by the rank of Σ. Σ^+ denotes the Moore-Penrose inverse of Σ. The p-value of the conditional test for H_0^j is then defined as:

$$P_j = \mathbb{P}_{H_0^j}(c(\mathbf{T}_j(\mathcal{L}_n, \mathbf{w}), \mu_j, \Sigma_j) \geq c(\mathbf{t}_j, \mu_j, \Sigma_j) \,|\, S(\mathcal{L}_n, \mathbf{w})) \tag{6.8}$$

The covariate X_{j^*} with the minimum p-value is then chosen.

The second step, denoting the binary split for this selected variable, is also implemented in the permutation test framework. The linear statistic for all possible subsets A of the sample space $\mathcal{X}_{j^*}$ is defined as follows:

$$\mathbf{T}_{j^*}^A(\mathcal{L}_n, \mathbf{w}) = \text{vec}\left(\sum_{i=1}^{n} w_i I(X_{j^*i} \in A) h(Y_i, (Y_1, \dots, Y_n))^\top\right) \in \mathbb{R}^q \tag{6.9}$$

The best split A^* which maximizes a test statistic over all possible subsets A is then chosen:

$$A^* = \arg\max_A c(\mathbf{t}_{j^*}^A, \mu_{j^*}^A, \Sigma_{j^*}^A) \tag{6.10}$$

For the classification case, the influence functions are defined with $h(Y_i, (Y_1, \dots, Y_n)) = e_J(Y_i)$ for the levels $1, \dots, J$. The conditional class probabilities for $y = 1, \dots, J$ are estimated with:

$$\hat{\mathbb{P}}(Y = y \mid X = x) = \left(\sum_{i=1}^{n} w_i(x)\right)^{-1} \sum_{i=1}^{n} w_i(x) I(Y_i = y) \tag{6.11}$$

Since in this credit scoring case missing values are not critical, Hothorn et al. (2006) is referenced for the handling of missing values.

6.2 Model-Based Recursive Partitioning

The so-called model-based recursive partitioning is another variant of recursive partitioning algorithms. The following descriptions for this method are based on the approach of Zeileis et al. (2008).

Basis for the description denotes a parametric model $\mathcal{M}(Y, \theta)$ with observations $Y \in \mathcal{Y}$ and a vector of parameters $\theta \in \Theta$. For fitting the model and receiving a parameter estimate $\hat{\theta}$, the objective function $\Psi(Y, \theta)$ should be minimized

$$\hat{\theta} = \underset{\theta \in \Theta}{\arg\min} \sum_{i=1}^{n} \Psi(Y_i, \theta) \tag{6.12}$$

If Ψ represents the negative log-likelihood, this leads to the well-known maximum likelihood estimator described in Section 3.1.

In the recursive partitioning case, the idea is not to estimate a single global model for all observations, but partition the observations according to some covariates and fit models in each cell of the partition. The aim is to grow a tree in which each node is associated with a model of the type $\mathcal{M}$. By evaluating a fluctuation test for parameter instability, it is tested as to whether a node should be divided further. Therefore, a node is split, if significant instability arise concerning any of the partitioning variables. The procedure is repeated until no significant instabilities are found any more and results in a tree with models of the type $\mathcal{M}(Y, \theta)$ in the terminal nodes. The algorithm can be described in four steps:

1. By the minimization of the objective function Ψ and the estimation of $\hat{\theta}$, the model is fitted once to all observations in the current node.

2. The stability of the parameter estimates is assessed with respect to every partitioning variable $Z_1, \dots, Z_l$. If this is stable, the procedure is stopped. Otherwise, if there is some overall instability, the variable Z_j is chosen with the highest parameter instability.

3. The split point(s) are computed that locally optimize the objective function Ψ.

4. The node is split into daughter nodes and the procedure is repeated.

The first step is common practice. It can be shown that $\hat{\theta}$ is received by solving the first order conditions with

$$\sum_{i=1}^{n} \psi(Y_i, \hat{\theta}) = 0 \qquad (6.13)$$

where $\psi(Y, \theta) = \frac{\partial \Psi(Y,\theta)}{\partial \theta}$ is the score function concerning $\Psi(Y, \theta)$. For a great variety of models, fitting algorithms are available for estimating $\hat{\theta}$; for instance, the maximum likelihood estimation with iterative weighted least squares.

In the second step, the aim is to analyze the stability of the parameter estimations concerning each of the partitioning variables Z_j and where required, to improve the fit by splitting the sample and capturing the instabilities. The tests used in this context result from the framework of generalized M-fluctuation tests presented in Zeileis and Hornik (2007).

The score function evaluated at the estimated parameters $\hat{\psi}_i = (Y_i, \hat{\theta})$ is used for testing the instability. This is analyzed with the scores $\hat{\psi}_i$ assessing whether they fluctuate randomly around their mean 0 or showing systematic deviations from the mean 0. These deviations can be analyzed with the following empirical fluctuation process

$$W_j(t) = \hat{J}^{-1/2} n^{-1/2} \sum_{i=1}^{\lfloor nt \rfloor} \hat{\psi}_{\sigma}(Z_{ij}) \qquad (0 \leq t \leq 1) \qquad (6.14)$$

where $\sigma(Z_{ij})$ is the ordering permutation defining the antirank of the observations Z_{ij} in the vector $Z_j = (Z_{1j}, \ldots, Z_{nj})^\top$ and $\hat{J}$ denotes an estimate of the covariance matrix $\text{cov}(\psi(Y, \hat{\theta}))$. $W_j(t)$ represents therefore the partial sum process of the scores ordered by the variable Z_j and scaled by the number of observations n.

As a test statistic for numerical partitioning variables Z_j, the $\sup LM$ statistic is used

$$\lambda_{\sup LM}(W_j) = \max_{i=\underline{i},\ldots,\overline{i}} \left(\frac{i}{n} \cdot \frac{n-i}{n} \right)^{-1} \left\| W_j \left(\frac{i}{n} \right) \right\|_2^2 \qquad (6.15)$$

which denotes the maximum over all single split LM statistics with a minimal segment size $\underline{i}$ and $\overline{i} = n - \underline{i}$. The corresponding p-value p_j can be computed from the supremum of a squared, k-dimensional tied-down Bessel process $\sup_t (t(1-t))^{-1} \|W^0(t)\|_2^2$ (Hansen, 1997). To analyze the instability for categorical variables Z_j with the categories C, the following test statistic is given

$$\lambda_{\chi^2}(W_j) = \sum_{c=1}^{C} \frac{|I_c|^{-1}}{n} \left\| \Delta_{I_c} W_j \left(\frac{i}{n} \right) \right\|_2^2 \qquad (6.16)$$

which is insensitive to the ordering of the category levels. $\Delta_{I_c} W_j$ denotes the sum of the scores in category c represented as the increment of the empirical fluctuation process over the observations in category $c = 1, \ldots, C$ (with the indexes I_c). The test statistic is χ^2 distributed and the related p-values p_j can be computed with $k \cdot (C-1)$ degrees of freedom

(Hjort and Koning, 2002). The statistic denotes the weighted sum of the squared L_2 norm of the increments.

To test for overall instability, it has to be assessed if the minimal p-value $p_{j^*} = \min_{j=1,\dots,l} p_j$ falls below a specified significance level α. The Bonferroni adjustment can be used to adjust for multiple testing. If significant instability arise, the variable Z_{j^*} with the minimal p-value is selected to divide the node further.

In the third step, the chosen variable Z_{j^*} is used to split the observations in the node in B child nodes. For each potential split point, the objective functions Ψ can be determined for the fitted models in the different B segments. In an exhaustive search, the split point with the minimal value of the objective function is chosen. In the credit scoring case, it is concentrated on binary splits with $B = 2$.

This ends one iteration of the algorithm, and then the procedure is repeated in the arising child nodes until no more significant instability is measured. Figure 6.4 shows a logistic-regression-based tree for the exemplary credit scoring data set with 1000 observations.

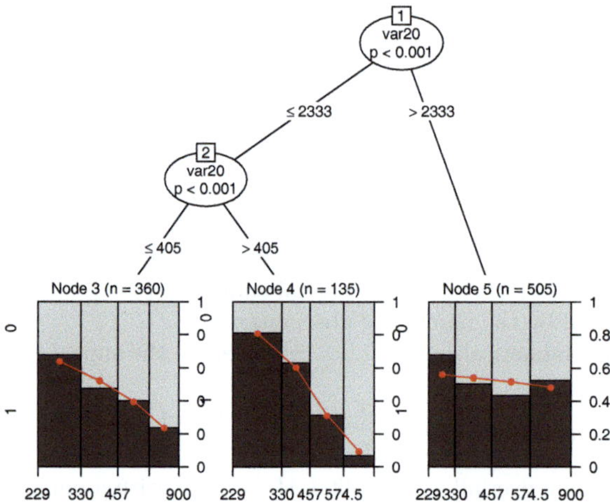

Figure 6.4: Example of a logistic-regression-based tree on credit scoring data for 1000 observations with var20 as partitioning variable, var03 as regressor and the binary dependent variable including non-defaults (0) and defaults (1). The default rate denotes 50%.

In the first iteration, instability existed leading to the split point with the value 2333 of variable 20. The right node then results in a logistic regression model with 505 observations, while the left node is divided further. The tests for instability described above lead to two further child nodes. The arising left node contains 360 observations having smaller values or equal 405 for the variable 20. The right node includes 135 observations greater than 405, resulting in a logistic regression model. For exemplary reasons, the model is created with one regressor (var03) and one potential partitioning variable with the covariate var20. Therefore, a spinogram visualizes the models in the terminal nodes by plotting the dependent binary variable against the numerical regressor var03. The binary outcome still denotes 0 for non-defaults and 1 for defaults. The breaks in the spinograms are defined by the five-point summary, while the fitted lines denote the mean predicted probabilities in each group. The grey shaded areas represent the relative frequencies of non-defaults and defaults in the groups. In general, the model tree in this exemplary credit scoring case determines three different groups.

6.3 Random Forests

Random Forests belong to the ensemble learning methods, which combine a whole set of single classification and regression trees. A main disadvantage of single trees is the instability to small changes in the learning data. The splitting decisions made at the beginning of the algorithm determine the splitting rules in the arising child nodes and depend mainly on the learning sample. Predictions of single trees can therefore show high variability. To overcome these problems, ensemble methods use the average over an ensemble of trees for prediction. The performance can then be improved due to a variance reduction of the predictions.

Breiman (1996) propose bagging (bootstrap aggregating) as advancement to the tree methodology. Later on Breiman (2001a) proposes random forests, which add an additional effect of randomness to bagging. In both algorithms, bootstrap samples or subsamples are drawn randomly from the learning sample and an individual tree is grown on each sample. Unpruned trees are used for the methods, i.e., the trees are grown very large without any stopping (Strobl et al., 2009). Random forests are one of the most popular machine learning algorithms, especially in scientific fields like genetics and bioinformatics. In other scientific fields as well, a huge variety of publications arose in the last few years. The applications, evaluations and developments in the financial field regarding the predictive accuracy are, however, rather rare.

6.3.1 Random Forest Algorithm

The random forest algorithm can be described following the descriptions of Liaw and Wiener (2002):

Random Forest Algorithm

1. Draw *ntree* bootstrap samples from the learning sample. *ntree* is the parameter of the random forest algorithms describing the number of trees.

2. For each of this generated bootstrap samples, grow a classification (or regression) tree regarding the following: in each node not all predictor variables are used to choose the best split but a special number of randomly selected predictors. The parameter *mtry* describes this number of randomly preselected splitting variables in the random forest algorithm. As well as trees in bagging, random forests grow unpruned trees though they are grown very large without any stopping (Strobl et al., 2009). When *mtry* equals the number of predictors, bagging can be thought of as the special case of random forests.

3. For predicting new data the aggregation of the predictions from the *ntree* trees is needed, i.e. majority votes for classification, average for regression.

Random forests produce a very diverse set of trees due to the specification of the potential covariates included in the algorithm. Bagging represents a special case in random forests when the number of randomly preselected splitting variables equals the overall number of covariates. In Breiman (2001a) it is stated that reducing the correlation between the single trees can improve the performance by a low upper bound for the generalization error. Hard decision boundaries are smoothed out with random forests. Additionally, due to the algorithm procedure, less important covariates can be included in the ensemble when stronger variables are not randomly selected in an iteration (Strobl et al., 2009). Therefore, interactions can be covered, which would not have been detected without including these covariates.

Different views exist about the number of trees that should be growing in random forests. Breiman (2001a) states in his work that random forests do not overfit and improve predictive accuracy with a low value of the generalization error. On the contrary, Lin and Jeon (2006) show, for example, that growing unpruned and therefore very large trees do not inevitably produce the best result. According to them, it depends on the situation. In cases with small sample sizes and high-dimensional data, growing large trees would be the best strategy. In other situations, it is preferable to tune the tree size for optimal results.

To construct random forests following the original approach of Breiman, the tree algorithm described in Section 6.1.1 can be used. In addition, the conditional inference framework described in Section 6.1.2 can also be used to create random forests by maintaining the rationale of Breimans approach. In combination with subsampling without replacement instead of bootstrap sampling, the latter offers unbiased variable selection and variable importance measures (Strobl et al., 2007). The tree algorithms are binary and recursive and represent a non-parametric approach without assuming a special distribution.

6.3.2 Importance Measures

Random forests are not only relevant for improving the predictive accuracy, but also present techniques for analyzing the variable importance to show the relevance of the different covariates.

A simple measure is to count the variables used in the forest. This is an easy approach to see the number of times a covariate is used for the splitting process. However, it is not considered in which position the variable is used for the split within the tree and the discriminatory power is not regarded.

In the original algorithm, Breiman (2002) describes different measures for evaluating the importance of the variables in the classification. Here, two popular importance measures are presented :

1. **Mean decrease in node impurity**: This measure implies the reduction of the node impurity measure calculated by the Gini index for classification. In the case of regression, it is measured by the residual sum of squares. The total decrease in node impurities from splitting on the variable is averaged over all trees.

2. **Mean decrease in accuracy**: This measure uses the so-called out-of-bag data (OOB) for computing. The idea is to randomly permute the predictor variable to cut the association with the response variable (Strobl et al., 2009). While drawing the bootstrap samples during the algorithm, two-thirds of the data are used. The remaining observations represent the out-of-bag data. The prediction error is recorded on this data before and after permuting the predictor variable. The difference between the two is averaged over all trees and normalized by the standard deviation of the differences.

Since the interpretation of random forests is more complicated and hard to visualize compared to single classification trees, both measures are reasonable indicators for analyzing the structure and the importance of variables within a forest. Nevertheless, in the original framework both measures are biased in the way that the variable selection favors covariates with many categories (Hothorn et al., 2006). Therefore, the permutation accuracy importance, as the most advanced variable importance measure, is embedded in the conditional inference framework. For a comparison and simulation studies to the bias in random forest variable importance measures, see the explanations of Strobl et al. (2007).

As described above, the rationale of the permutation accuracy importance is that the association between the variable and the response should be broken by randomly permuting the values of a covariate (Strobl et al., 2009). The difference of the predictive accuracy before and after permuting the variable averaged over all trees, can then be used as variable importance measure.

For the following explanations compare Strobl et al. (2008) and Strobl and Zeileis (2008). The importance of variable X_j in tree t is formally defined as

$$VI^{(t)}(X_j) = \frac{\sum_{i \in \overline{\mathfrak{B}}^{(t)}} I\left(y_i = \hat{y}_i^{(t)}\right)}{\left|\overline{\mathfrak{B}}^{(t)}\right|} - \frac{\sum_{i \in \overline{\mathfrak{B}}^{(t)}} I\left(y_i = \hat{y}_{i,\psi_j}^{(t)}\right)}{\left|\overline{\mathfrak{B}}^{(t)}\right|} \tag{6.17}$$

where $\overline{\mathfrak{B}}^{(t)}$ denotes the out-of-bag sample for a tree t with $t \in \{1, \ldots, \text{ntree}\}$. $\hat{y}_i^{(t)} = f^{(t)}(x_i)$ defines the predicted class for observation i before permuting its value of variable X_j while $\hat{y}_{i,\psi_j}^{(t)} = f^{(t)}(x_{i,\psi_j})$ is the predicted class for observation i after permuting. If a variable X_j is not included in a tree t, the variable importance is defined as $VI^{(t)}(X_j) = 0$. Then, the average importance over all trees denotes the raw importance score for each covariate

$$VI(X_j) = \frac{\sum_{t=1}^{ntree} VI^{(t)}(X_j)}{ntree} \tag{6.18}$$

By dividing the raw importance measure with the standard error, the scaled version of the importance measure, the so-called z-Score

$$z(X_j) = \frac{VI(X_j)}{\frac{\hat{\sigma}}{\sqrt{ntree}}} \tag{6.19}$$

By using subsampling without replacement for building the forest, combined with conditional inference trees, the permutation importance is unbiased and reliable, even if the covariates do not have the same scale. A strong association between the covariate and the response is indicated with a large value of the permutation importance, while values around zero or even negative values represent weak relation (Strobl et al., 2009). The fact that the importance measure considers the influence of a variable individually, as well as in interaction with other covariates, represents the main advantage of the measure.

The corresponding null hypothesis of independence for the original permutation importance shows, that the importance measure not only covers the relation between the response Y and the variable X_j, but also reflects, due to the structure of a tree, the remaining variables $Z = X_1, \ldots, X_{j-1}, X_{j+1}, \ldots, X_p$. Therefore, the null hypothesis for permuting X_j against the outcome Y results in

$$H_0 \colon X_j \perp Y, Z \text{ or equivalently } X_j \perp Y \wedge X_j \perp Z \tag{6.20}$$

Obviously, this implies that high importance measures can result from the relation between X_j to Y, and also from the relation to Z. For that reason, Strobl et al. (2008) propose a conditional permutation scheme, which takes the correlation structure between X_j and the other covariates into account. The corresponding null hypothesis

$$H_0 \colon (X_j \perp Y) \mid Z \tag{6.21}$$

reflects the permutation of X_j in dependence of Z. In the case where X_j and Z are independent, both measures have the same result.

Which permutation scheme should be used depends on the specific research question. The conditional importance measure is computational very intensive and can therefore not be computed for the presented credit scoring case. Therefore, I concentrate on the unconditional version and another variant of the variable importance measure that is based on the AUC measure. Janitza et al. (2013) propose a novel AUC-based permutation variable importance measure for random forests that is more robust in cases with unbalanced data, i.e., where one response class is more represented than the other one. The rationale of this approach is very similar to the standard accuracy variable importance measure, but differs in the predictive accuracy measure. Instead of using the error rate, the AUC is implemented as measure before and after permuting a covariate. The definition of the AUC-based variable importance for variable j follows

$$VI_j^{(AUC)} = \frac{1}{ntree^*} \sum_{t=1}^{ntree^*} \left(AUC_{tj} - AUC_{t\tilde{j}}\right) \tag{6.22}$$

where $ntree^*$ indicates the number of trees in the forest where both classes are represented in the OOB data. While AUC_{tj} denotes the AUC measure in tree t computed on the OOB data before randomly permuting variable j, $AUC_{t\tilde{j}}$ is the AUC measure after permuting covariate j.

The various variable importance measures for random forests are evaluated and compared for the credit scoring case in Section 6.4.3.

6.4 Recursive Partitioning Methods for Credit Scoring

Before beginning the evaluation of the recursive partitioning methods for the presented credit scoring case, some related work in the academic literature is emphasized for retail scoring models. Besides the classical scoring models like logistic regression or discriminant analysis, most studies in previous years evaluate methods like neural networks and support vector machines. Crook et al. (2007) or Lessmann et al. (2013) give an overview of recent surveys. Most analyses conclude in terms of classification trees lower performance compared to other classifiers in the credit scoring context. However, there also exist surveys that indicate higher predictive accuracy for classification trees. Single studies outline the good performance of random forests for credit scoring models.

In the evaluation of Baesens et al. (2003), C4.5 rules perform poorer statistically than other classifiers like neural networks or logistic regression. Brown and Mues (2012) analyze different classifiers with an evaluation on five different credit scoring data sets. The applied C4.5 decision trees show inferior results compared to the other algorithms, while random forests and gradient boosting indicate the best results within this survey. The authors, however, note that the classical logistic regression is still competitive for their presented evaluation. In Lee et al. (2006) and Xiao et al. (2006), superior outcomes are shown for CART algorithms and multivariate adaptive regression splines (MARS) (cf. Friedman (1991)) compared to logistic regression, neural networks and support vector machines. Since the former study is based on credit card data from Taiwan, the latter analysis relies on public German and Australian credit data sets. In Lessmann et al. (2013) and Kruppa et al. (2013), random forests are indicated as best classifiers in the evaluation.

Note that different measures, like the misclassification rate, are used for comparison reasons in the presented studies and many surveys are based on often-used public data sets. Moreover, CART and C4.5 are presented as classification trees and the classical random forest algorithm as representative for this method. In the evaluation, conditional inference trees are additionally presented that overcome some weaknesses of the classical algorithms. Furthermore, random forests based on the classical CART and conditional inference trees are analyzed. In Section 6.4.2 a combination of recursive partitioning and logistic regression is investigated for the credit scoring data.

6.4.1 Classification Trees for Credit Scoring

CART-like trees. To begin the evaluation of classification trees in credit scoring, the approach of Breiman's CART algorithm is analyzed that is explained in Section 6.1.1. The main tuning parameters are

- minsplit - denotes the minimal number of observations in a node to divide this node further
- cp - denotes the cost complexity parameter α described in Section 6.1.1

The procedure for training the data with classification trees and tuning the parameters is described for the 50%-training sample. Instead of using the missclassification error, the AUC measure is used for choosing the parameters. The Gini index is used as split criterion for the example, and the results are presented in Table 6.1. First of all, for a fixed minsplit value, a variety of different cost complexity values is measured by cross-validation (cf. for details Therneau and Atkinson (1997)). Since the predictive accuracy is examined for each subtree, the tree with the best AUC value is chosen. The first square in Table 6.1, therefore, shows the tree with the best AUC value for a special minsplit value of a sequence of trees with different cp-values. The cost-complexity value is relevant for pruning the tree to avoid overfitting. A value of 0 for α creates full trees, while large values of α lead to smaller trees. The number of splits is contained in the table and indicates the tree size.

AUC	lowCI95%	upCI95%	minsplit	cp-value	number of splits
0.6686	0.6585	0.6787	10	0.00395778	10
0.6698	0.6597	0.6799	50	0.00329815	22
0.6749	0.6648	0.6849	100	0.00065963	20
0.6754	0.6654	0.6855	150	0.00329815	12
0.6476	0.6374	0.6577	200	0.00527705	7
0.6476	0.6374	0.6577	250	0.00527705	7
...	...	...	...	...	...
0.6452	0.6350	0.6553	1000	0.01418206	1
0.6698	0.6597	0.6799	50	0.00329815	22
0.6713	0.6612	0.6814	60	0.00109938	45
...	...	...	...	...	...
0.6750	0.6650	0.6851	110	0.00039578	23
0.6761	0.6660	0.6861	120	0.00065963	19
0.6754	0.6654	0.6855	130	0.00329815	12
0.6754	0.6654	0.6855	140	0.00329815	12
0.6754	0.6654	0.6855	150	0.00329815	12
0.6750	0.6650	0.6851	110	0.00039578	23
0.6750	0.6650	0.6851	111	0.00039578	23
...	...	...	...	...	...
0.6745	0.6645	0.6846	115	0.00039578	20
0.6767	0.6666	0.6867	116	0.00039578	22
...	...	...	...	...	...
0.6754	0.6654	0.6855	130	0.00329815	12

Table 6.1: Prediction accuracy for CART classification trees with different complexity parameters (cp-values) and minsplit values, trained on the 50%-training sample and applied to the test sample with the Gini index as split criterion and continuous covariates. The number of splits indicates the tree size.

Because there are too many combinations to examine for different tuning parameters, the minsplit values are increased in steps of 50 (except the first value of 10). It is analyzed which interval yields the best predictive performance. In the example shown in Table 6.1, this is the interval of 50 to 150. In a next step, the same procedure with a variety of cp-values is applied to different minsplit values with intervals of 10. The results are shown in the middle square of Table 6.1.

This procedure continues with the minsplit values between 110 and 130 shown in the bottom square of Table 6.1. For the presented case, the results are similar for small changes in the minsplit value. The algorithm is trained on the training sample, while the tuning parameters are optimized according to the outcomes on the test sample. The validation sample is used to prove the findings.

The best result for training the 50%-training data yields an AUC value of 0.6767 on the test sample with a minsplit value of 116. This implies that the predictive accuracy for the classification tree lies below the AUC measure of the logit model with 0.7051.

The procedure described above is applied on the three different training samples. The minsplit value and the cp-value are tested with the Gini index as split criterion and tuned on the test sample. Table 6.2 shows the best results of the huge amount of different combinations according to the AUC measure. The predictive accuracy for the logit model is included to compare the results, and the number of splits shows the size of the trees.

	AUC	minsplit	cp-value	number of splits	split criterion	AUC LogReg
trainORG	0.6573	15	0.00012270	169	gini	0.7045
train020	0.6544	35	0.00122699	75	gini	0.7054
train050	0.6767	116	0.00039578	22	gini	0.7051

Table 6.2: Prediction accuracy for the best CART classification trees trained on the three different training samples with continuous covariates, applied to the test sample with the tuning parameters minsplit, cp-value and the Gini index as split criterion.

Even if these are the best results out of the whole procedure of tuning the different parameters and testing a lot of different combinations, the predictive accuracy of the classification trees is below the AUC values of the logit model. For the original training sample, the tree is relatively large with 169 splits and a small cost complexity parameter of 0.00012270. The corresponding AUC for the classification tree underachieves with 0.6573 compared to 0.7045 of the logit model. This tree includes 20 covariates for the tree construction. As already described above, the tuning parameter α covers the tradeoff between the tree size and the goodness of fit to the data. For the 20%-training sample, the complexity parameter indicates a higher value and presents a smaller tree size with 75 splits. The model for this training sample uses 19 variables for the tree construction. The classification tree trained on the 50%-training sample yields an AUC value of 0.6767 on the test data with 11 covariates used for building the tree and a relatively small tree size of 22 splits. The outcomes of all three samples show that the classification tree algorithm cannot outperform the classical logistic regression model.

The results on the validation sample approve the findings. Table 6.3 shows the results with the optimized tuning parameters applied to the validation sample. For the presented credit scoring case, the classification trees based on the CART algorithm indicate low performance compared to the logit model on different performance measures. The AUC differences on the validation data are significant according to DeLong's test. Since the MER values are on the same level, all other measures show better performance for the logistic regression results.

		AUC	*p*-value	H measure	KS	MER	Brier score
trainORG	CART	0.6504		0.0840	0.2346	0.0272	0.0301
	LogReg	0.7118	0.0000	0.1283	0.3101	0.0272	0.0270
train020	CART	0.6624		0.0883	0.2863	0.0272	0.0783
	LogReg	0.7129	0.0000	0.1286	0.3143	0.0272	0.0719
train050	CART	0.6814		0.0906	0.2827	0.0272	0.2577
	LogReg	0.7125	0.0002	0.1276	0.3142	0.0272	0.2494

Table 6.3: Validation results for the best CART classification trees trained on the different training samples and applied to the validation sample measured with the AUC. The related minsplit, cp-values and number of splits are shown in Table 6.2. The *p*-values result from DeLong's test.

As an impurity measure, it is concentrated on the Gini index explained in Section 6.1.1. However, the Information index, measuring the entropy also explained in Section 6.1.1, is tested for the 50% training sample. The best results of the evaluation procedure concerning the AUC are contained in Table A.2 in the Appendix. The AUC measures slightly underachieve the results produced with the Gini index as impurity measure. Therefore, the Gini index is used for further evaluation of the classification trees based on CART.

For the original training sample with a low default rate, altering the prior probabilities was also tested. But the AUC measure on this credit scoring data can not be improved by using different priors (cf. Table A.3 in the Appendix). Note that altered priors only affect the choice of split, while the original priors are used to compute the risk of the node (Therneau and Atkinson, 1997). In addition, the two additional training samples with high default rates of 20% and 50% are evaluated.

As described in Section 6.1.1, it is common to improve the probability estimates in classification trees with smoothing techniques in order to improve the predictive accuracy. The Laplace smoothing as well as the m-estimate smoothing is applied for the presented credit scoring case. For the former case, the procedure described above was applied to find the best pruned tree with the best minsplit values. For the Laplace smoothing, the parameters are $m = 2$ and $p = 0.5$. Table 6.4 shows the best results from the entire process of evaluating different minsplit and cp-values.

	AUC	minsplit	cp-value	number of splits	AUC LogReg
trainORG	0.6764	20	0.00010225	134	0.7045
train020	0.6642	23	0.00046012	165	0.7054
train050	0.6761	116	0.00065963	19	0.7051

Table 6.4: Prediction accuracy for the best CART classification trees with Laplace smoothing and the gini split criterion, trained on the different training samples and applied to the test sample.

The outcomes are trained on the different training samples and applied to the test sample for optimizing the parameters regarding the AUC. Compared to the AUC values in Table 6.2 estimated with relative frequencies, the predictive accuracy can be improved with Laplace smoothing. Two samples show higher AUC values for the classification trees with Laplace smoothing. However, the comparison to the logit model still indicates poor results for the classification trees. For instance, an AUC value of 0.7045 for the logit model on the test sample, trained on the original training sample denotes higher predictive accuracy than the corresponding AUC measure of 0.6764 for the classification tree with Laplace smoothing.

The same conclusions can be drawn from the validation sample in Table 6.5. The results with Laplace smoothing outperform the classification trees with relative frequencies, but still yield poor results compared to the classical logistic regression model. The different performance measures approve the previous findings on the validation data and show higher performance for the logit model.

		AUC	*p*-value	H measure	KS	MER	Brier score
trainORG	CART	0.6731		0.0927	0.2670	0.0272	0.0290
	LogReg	0.7118	0.0000	0.1283	0.3101	0.0272	0.0270
train020	CART	0.6639		0.0814	0.2681	0.0272	0.0915
	LogReg	0.7129	0.0000	0.1286	0.3143	0.0272	0.0719
train050	CART	0.6845		0.0944	0.2717	0.0272	0.2461
	LogReg	0.7125	0.0009	0.1276	0.3142	0.0272	0.2494

Table 6.5: Validation results for the best CART classification trees with Laplace smoothing and the gini split criterion, trained on the different training samples and applied to the validation sample. The related minsplit, cp-values and number of splits are shown in Table 6.4. The *p*-values result from DeLong's test.

The m-estimate smoothing is also applied for the credit scoring case explained in Section 6.1.1. In addition to find the optimal minsplit and cp-values regarding the AUC measure, the optimal parameters for m are included in the optimization procedure. The probability parameter p are set to 0.974, which denotes the overall probability of a non-default. For the parameter m, values of $5, 10, 20, 50, 100, 500, 1000$ are tested. In order to reduce the huge amount of different combinations, I limited the minsplit values to groups of ten. The results are trained on all three training samples, while the parameters were optimized on the test sample. Table A.4 in the Appendix shows the best results from a large variety of different outcomes. The validation results are included in Table A.5 in the Appendix.

Since the Laplace smoothing could improve the predictive accuracy for the classification trees compared to the classical CART method, the m-estimate smoothing also improves the results with relative frequencies. Compared to the Laplace smoothing, the AUC values for the classification trees are better in some cases, while in other cases the m-estimate smoothing outperforms. Both probability smoothing techniques for classification trees yield AUC measures below the logit model predictive accuracy.

Conditional inference trees. The evaluation of the credit scoring data is continued with the conditional inference trees described in Section 6.1.2. The former tree algorithms are biased in favor of continuous covariates and variables with many potential cutpoints. Conditional inference trees overcome this problem by ensuring unbiased variable selection (cf. the descriptions in Section 6.1.2). Additionally, the statistical approach of the algorithm indicates that no form of pruning or cross-validation is needed.

Therefore, the parameter minsplit is tuned on the training and test data following the procedure described for the CART-like trees. Multiplicity Bonferroni-adjusted p-values are used for the stopping criterion. The criterion is maximized meaning that the $1 - p$-value is used. A node is split if the value of the so-called mincriterion is exceeded. For the presented case, the mincriterion is set to 0.95, i.e., the p-value must be smaller than 0.05 for splitting the node further.

Additionally, two variables had to be adjusted in the way that a few negative values were set to zero. This results from a technical effect in the application process, that the variables slightly changed over time what the presented algorithm could not handle. The effect is rather small but guarantees the inclusion of all 26 variables in the evaluation process. Moreover, data sets with encoded meaning were excluded from the samples in order to use continous variables.

Table 6.6 shows the best outcomes of the optimization concerning the AUC measure. Continuous covariates were used for this analysis. The results on the test sample as well as on the validation sample are contained.

	AUC	lowCI95%	upCI95%	minsplit	AUC LogReg	p-value	
trainORG	0.6897	0.6797	0.6997	488	0.7045		test
train020	0.6843	0.6743	0.6943	324	0.7054		test
train050	0.6683	0.6582	0.6784	10	0.7051		test
trainORG	0.6866	0.6728	0.7005	488	0.7118	0.0029	validation
train020	0.6837	0.6699	0.6976	324	0.7129	0.0006	validation
train050	0.6756	0.6617	0.6895	10	0.7125	0.0000	validation

Table 6.6: Prediction accuracy for the best conditional inference trees trained on the different training samples with continuous covariates and applied to the test and validation samples. The p-values result from DeLong's test.

Obviously, even if the classification algorithm is optimized regarding the unbiased variable selection, the predictive accuracy lies definitely below the AUC values of the logit model for all three training data sets on the test sample. For instance, the results for the original training sample applied to the test sample with a minsplit value of 488, yields an AUC of 0.6897 for the conditional inference trees, compared to 0.7045 for the logit model. This conditional inference tree has 33 terminal nodes and includes 12 covariates for the tree building. The model trained on the 20% training sample yields an AUC value of 0.6843 on the test data and includes 13 variables within the tree with 23 terminal nodes. For the

50% training data, the best result is yielded with a minsplit value of 10 and an AUC value of 0.6683 compared to 0.7051 of the logit model. The conditional inference tree trained on this data denotes 14 terminal nodes with 7 variables used for the tree. The covariates used for the conditional inference trees include the most important variables from the logit model. For instance, variable 3 and 4 are included in the estimated classification trees.

The outcomes on the validation data confirm the findings for all three samples. The differences of the AUC values between the conditional inference trees and the logit models on the validation are significant according to DeLong's test. While the conditional inference tree trained on the original training sample yields 0.6866 on the validation data, the logit model shows an AUC value of 0.7118.

This difference is apparent in the ROC-graphs displayed for the validation data trained on the original training sample. Figure 6.5 includes the two graphs, which show that the curve for the conditional inference tree is definitely below the one of the logistic regression model (dashed line).

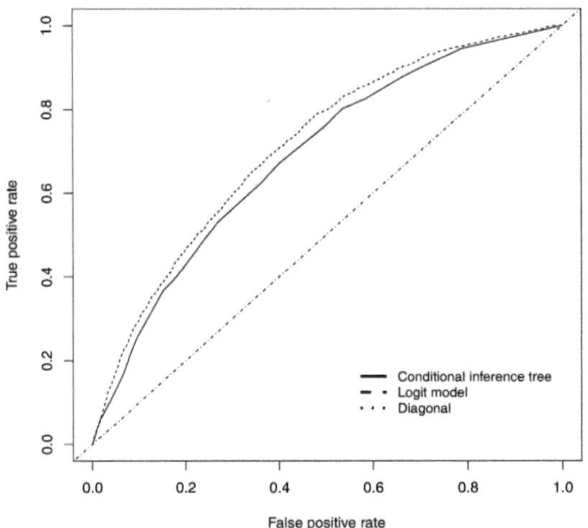

Figure 6.5: ROC graphs for the conditional inference tree (AUC 0.6866) and the logit model (AUC 0.7118) trained on original training sample applied to the validation data.

Different performance measures for the validation results are shown in Table 6.7. The MER values indicate similar performance for the conditional inference trees and the logit models. The Brier scores are better for the validation results trained on the 20% and 50% training samples. But the H measures and the KS values show higher performance for the logit model. The H measure for the conditional inference tree, trained on the original training sample, is 0.0994 compared to 0.1283 for the corresponding logit model. Similar results are indicated for the results trained on the other both training samples. The KS value denotes 0.2702 for the conditional inference tree trained on the original training sample and applied to the validation data. The corresponding value of the logit model is 0.3101.

		AUC	Gini	H measure	KS	MER	Brier score
trainORG	Tree	0.6866	0.3732	0.0994	0.2702	0.0272	0.0270
	LogReg	0.7118	0.4237	0.1283	0.3101	0.0272	0.0270
train020	Tree	0.6837	0.3675	0.0998	0.2879	0.0272	0.0676
	LogReg	0.7129	0.4257	0.1286	0.3143	0.0272	0.0719
train050	Tree	0.6756	0.3512	0.0826	0.2632	0.0272	0.2328
	LogReg	0.7125	0.4249	0.1276	0.3142	0.0272	0.2494

Table 6.7: Further measures for the conditional inference tree results with continuous variables compared to the logistic regression results for the validation data.

The results for the conditional inference trees with classified variables are presented in the following. Four of the covariates are original categorical, while the other 22 variables are included as categorized effects. The same procedure is applied as described above for the conditional inference trees. Since no form of pruning or cross-validation is needed for the presented approach, Table 6.8 shows the AUC values for the trees and the corresponding logit model as well as the minsplit values. The results are the best outcomes of a huge variety of different minsplit values. The AUC values are better than the results of the conditional inference trees with continuous covariates. But the predictive accuracy is still below the AUC measure of the logistic regression model. For instance, on the original training sample, the AUC value for the test data denotes 0.6954 compared to 0.7204 of the logit model. This model presents 23 terminal nodes. On the validation data, the same results are shown with an AUC value of 0.6945 for the tree compared to 0.7263 for the logit model. The differences of the AUC measures between the trees and the logit models are significant on the validation data according to DeLong's test.

	AUC	lowCI95%	upCI95%	minsplit	AUC LogReg	*p*-value	
trainORG	0.6954	0.6858	0.7051	615	0.7204		test
train020	0.6898	0.6801	0.6995	322	0.7211		test
train050	0.6860	0.6763	0.6957	53	0.7211		test
trainORG	0.6945	0.6810	0.7079	615	0.7263	0.0000	validation
train020	0.6823	0.6688	0.6958	322	0.7282	0.0000	validation
train050	0.6861	0.6726	0.6996	53	0.7275	0.0000	validation

Table 6.8: Prediction accuracy for the best conditional inference trees with classified covariates, trained on the different training samples and applied to the test and validation samples. The *p*-values result from DeLong's test.

Table 6.9 shows different performance measures for the validation results with coarse classification. The MER values are equivalent for the conditional inference trees and the logit models. The Brier score shows better performance for the trees trained on the 20% and 50% training sample. All other performance measures indicate definitely better performance for the logit model. For instance, the H measure on the validation data, trained on the original training sample, denotes 0.1071 for the conditional inference tree compared to 0.1468 for the logit model.

		AUC	Gini	H measure	KS	MER	Brier score
trainORG	Tree	0.6945	0.3890	0.1071	0.2779	0.0260	0.0260
	LogReg	0.7263	0.4527	0.1468	0.3310	0.0260	0.0260
train020	Tree	0.6823	0.3645	0.0967	0.2547	0.0260	0.0574
	LogReg	0.7282	0.4564	0.1476	0.3314	0.0260	0.0658
train050	Tree	0.6861	0.3721	0.0971	0.2797	0.0260	0.2063
	LogReg	0.7275	0.4550	0.1473	0.3364	0.0260	0.2217

Table 6.9: Further measures for the conditional inference tree results with classified variables compared to the logistic regression results for the validation data.

The results for the evaluation with classified variables are similar to the results with continuous covariates. The same conclusions can be drawn.

In general, the various results for classification trees do not outperform the classical way of scorecard development for the presented credit scoring case. The advanced conditional inference trees have the advantage that no pruning in any form is needed and that the variable selection is unbiased. The results outperform the traditional CART-algorithm for the original training sample and the 20% training data applied to the test data. However, both algorithms show relatively poor results compared to the logit model. The good interpretability of classification trees due to their binary structure indicates an advantage of the classification approach.

6.4.2 Model-Based Recursive Partitioning in Credit Scoring

Another variant of recursive partitioning algorithms is the model-based recursive partitioning. The method is explained in Section 6.2 and is in the following evaluated for the credit scoring case.

The approach combines recursive partitioning with classical parametric model fitting. The model type is not restricted and can be specifically defined. Since the logit model is the basis and the benchmark in the current case, the focus lies on a logistic-regression based tree for the credit scoring data. A visualized example is given in Section 6.2 and displayed in Figure 6.4.

For analyzing the data, the instability is assessed using a Bonferroni-corrected significance level of $\alpha = 0.05$. The minimal segment size is used as tuning parameter. The basis for the evaluation are the 26 covariates. Since a large variety of combinations arise to analyze the variables as regression variables and partitioning variables, the procedure for analyzing the data is described in the following.

In Chapter 3, the classical way of scorecard development with logistic regression is described, resulting in a model with 8 explanatory variables. This model is used as basis for the presented approach of logistic-regression-based trees. First, each of these 8 most important variables is separately used as single covariate for the logit model. The remaining 25 variables are tested as partitioning variables. In a next step, the whole model with the 8 variables is used, while the remaining 18 covariates are tested as partitioning variables. According to these outcomes, special variable combinations are analyzed. For instance, the two most important covariates concerning the AUC are tested in the logistic model, while all other 24 variables are tested as partitioning variables. The AUC is used as a measure of performance for selecting the best models and for selecting the best value of the tuning parameter. For each variable combination, the minimal number of observations in a node is optimized concerning the AUC. The tuning parameter was set to $10, 100, 200, \ldots, 1000$, in a first step. Regarding the results, the values for the tuning parameter are refined in steps of 50 and finally in smaller steps of 10 and 1. Note that at least a few hundred observations should be included in the nodes to estimate the logit model. A minsplit value of 10 and 100 is therefore only analyzed for test reasons, and a minimal number of 200 observations in a node is used for comparing the results.

As already stated for the conditional inference trees, two variables had to be adjusted to enable the inclusion of all 26 variables. The adjustments are negligible and result from process related transformations. Furthermore, by using continuous explanatory variables, observations with encoded meaning had to be omitted (compare Section 4.3). The 50% training sample is used for training the model-based trees, the tuning parameter is optimized on the test data and the validation data are used to validate the outcomes.

I began the evaluation for the 50% training sample by including the 26 variables as continuous covariates. This means that the 22 numerical variables are not aggregated to classes. The four categorical effects, as explained in Section 2.2, remain categorical.

Table 6.10 shows the best results of the optimization process for the variables and the tuning parameter described above. The minimal number of observations in a node is set to

variable for the model	AUC	lowCI95%	upCI95%	minsplit	
var14	0.5517	0.5417	0.5618	200	test
var22	0.5542	0.5441	0.5643	200	test
var25	0.5701	0.5599	0.5802	200	test
var09	0.5906	0.5805	0.6008	200	test
var26	0.5993	0.5892	0.6095	200	test
var07	0.6189	0.6087	0.6290	467	test
var03	0.6411	0.6309	0.6512	394	test
var04	0.6619	0.6518	0.6720	400	test
var03 and var04	0.6710	0.6610	0.6811	370	test
8 variables	0.7098	0.6999	0.7197	400	test
var14	0.5412	0.5274	0.5551	200	validation
var22	0.5427	0.5288	0.5566	200	validation
var25	0.5747	0.5607	0.5887	200	validation
var09	0.5720	0.5580	0.5861	200	validation
var26	0.5938	0.5797	0.6078	200	validation
var07	0.6140	0.5999	0.6281	467	validation
var03	0.6309	0.6169	0.6450	394	validation
var04	0.6677	0.6538	0.6817	400	validation
var03 and var04	0.6733	0.6594	0.6873	370	validation
8 variables	0.7134	0.6998	0.7271	400	validation

Table 6.10: Results for the best logistic regression based trees trained on the 50% training samples, applied to the test and validation samples.

200 and the results showed that the algorithm, as expected, does not converge for small minsplit values. Moreover, the AUC values often equals for small changes in the minsplit value. For instance, if a minsplit value of 250 and 251 shows the same AUC measure, the smallest minsplit value is chosen.

Table 6.10 contains the results for the test sample on which the tuning parameter was optimized, and furthermore, contains the outcomes for the validation data. The indicated variables in the left column present the covariates for the logistic regression model. The first rows show therefore the results where only one potential covariate is included in the model and the remaining variables are tested as partitioning variables. All 8 variables of the logit model in Section 3.2 are analyzed as a single variable. Additionally, two, three and more variables are evaluated regarding the predictive accuracy. For instance, the results for variable 3 and variable 4 are shown in Table 6.10. This means that both variables are included in the logistic regression model whereby the stability of the parameter estimates is assessed with respect to all remaining 24 partitioning variables.

The best predictive accuracy of all tested variable combinations and optimized tuning parameters yields the logistic regression based tree with all 8 variables in the logit model. Since the AUC measure on the test data denotes, for example, 0.5517 for the model with variable 14 as model covariate, the logistic regression based tree with 8 variables yields 0.7098. In the latter algorithm, variable 15 is identified as variable with the smallest p-value among all other tested partitioning variables. The data is classified twice according to the

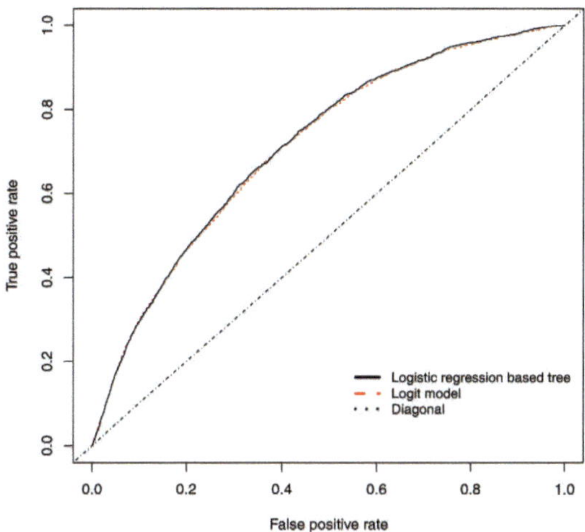

Figure 6.6: ROC graphs for the logistic regression based tree (AUC 0.7134) and the logit model (AUC 0.7125) trained on the 50% training sample applied to the validation data.

best split in variable 15 and results, therefore, in three regression models in the terminal nodes. The outcomes indicate that the improvement in the predictive accuracy is higher with more variables in the logit model than using them as partitioning variables. The analyses for the single classification trees in the former section already indicate that the predictive accuracy underachieves compared to the logit model. The combination of both methods, however, improves the performance.

The best result for the logistic regression based tree yields an AUC value of 0.7098 on the test data, while the corresponding AUC value on the same data for the logit model with unclassified covariates denotes 0.7051. On the validation data, the predictive accuracy for the recursive partitioning method is 0.7134 compared to 0.7125 for the logit model. Since there is only a slight improvement, the ROC-curves illustrated for the validation data in Figure 6.6, show almost the same trend. To highlight the difference, the dashed line for the logit model is red, indicating that it lies slightly beneath the ROC curve of the model based tree.

For the best model-based recursive partitioning tree with 8 variables and continuous covariates, different performance measures are shown in Table 6.11. The minsplit value for this model is 400. The comparison on the test data indicates better performance for the model-based tree for all different measures. But on the validation data, the H measure and the KS value denote slightly higher values for the logit model. The AUC difference is not significant according to DeLong's test.

		AUC	p-value	H measure	KS	MER	Brier score
test	model based tree	0.7098		0.1258	0.3159	0.0275	0.2373
	LogReg	0.7051		0.1211	0.3065	0.0275	0.2380
validation	model based tree	0.7134		0.1268	0.3139	0.0272	0.2473
	LogReg	0.7125	0.9080	0.1276	0.3142	0.0272	0.2494

Table 6.11: Different performance measures for the best logistic regression based tree with continuous variables and a minsplit value of 400 compared to the logistic regression results for the test and validation samples. 8 variables are used for the model. The p-value results from DeLong's test.

Additionally, the model-based recursive partitioning algorithm is tested for classified variables on the 50% training data. Four of the 26 explanatory variables are still categorical, and the other 22 variables are now included as categorical effects with different classes. The procedure is the same as described above and analyzed for the continuous variables. An extract of the results is given in Table 6.12.

The table shows four selected results, with one, two, three and eight variables in the logit model, while the remaining variables are used as partitioning variables. The best result is also yielded with the logit model including 8 variables and testing the stability of the parameter estimates with the other variables. The predictive accuracy is 0.7211 on the test data, compared to 0.7211 for the logit model. On the validation data, the results also equal an AUC value of 0.7275. For the model with 8 variables, the partitioning algorithm can not improve the predictive accuracy. The performance for the classified variables is, therefore, for both methods on the same level. For this reason, the different performance measures are not presented for this case. The tuning parameter minsplit was optimized concerning the AUC measure. The values in Table 6.12 show the best results of the optimization process.

variable for the model	AUC	lowCI95%	upCI95%	minsplit	
var04	0.6911	0.6814	0.7007	243	test
var04 and var03	0.6939	0.6843	0.7036	200	test
var04, var03 and var26	0.7018	0.6922	0.7114	200	test
8 variables	0.7211	0.7116	0.7306	389	test
var04	0.6979	0.6845	0.7113	243	validation
var04 and var03	0.6975	0.6841	0.7109	200	validation
var04, var03 and var26	0.7039	0.6905	0.7173	200	validation
8 variables	0.7275	0.7143	0.7406	389	validation

Table 6.12: Prediction accuracy for the best logistic regression based trees trained on the 50% training samples, applied to the test and validation samples with categorized covariates.

The model-based recursive partitioning algorithm improves the performance of the relatively poor performing single classification trees shown in Section 6.4.1. The combination of classification with classical parametric models is a reasonable approach that shows good performance for the credit scoring case. But the best results are still achieved with more variables in the logit model than using them as partitioning variables. Another disadvantage of logistic regression based trees is the computational intensity. Therefore, the evaluation is concentrated on the 50% training sample. The model-based classification can support the variable selection by testing the stability of the parameter estimates. An interesting way is to analyze the partitioning variables selected according to the smallest p-value for classification, not only as partitioning variables, but testing them in the logit model to compare the results.

6.4.3 Random Forests in Credit Scoring

Another recursive partitioning approach to overcome the instability of single classification trees are random forests, which combine a whole set of classification trees. The random forest algorithm is explained in Section 6.3.

Random forests with CART-like trees. The evaluation of random forests in credit scoring is started with the classical approach of Breiman (2001a) with CART-like trees. The variables are analyzed with the 'classical' importance measures 'mean decrease in node impurity' and 'mean decrease in accuracy'. The most important tuning parameters are the following:

- ntree - describes the number of individual trees that are grown during the process
- mtry - describes the number of randomly selected variables at each node

The training samples are used to train the random forests on the credit scoring data while the tuning parameters are tuned on the test sample. The results are validated on the validation sample to prove the findings. The AUC measure is used for analyzing and comparing the predictive accuracy of the models and is used specifically for tuning the parameters. The selection of the tuning parameters *ntree* and *mtry* follows, therefore, the level of the AUC measure. While using continuous covariates, data sets with encoded meaning were excluded from the data samples. Different performance measures are evaluated and compared for the final results.

The procedure for training the data and tuning the parameters can be described as follows. First of all, a specified number of trees is fixed to evaluate the best value for selecting the variables. According to the highest predictive accuracy, the best mtry-value is defined. Secondly, with this value for the mtry parameter, the number of trees is varied. For the unpruned trees within the random forest algorithm, the number of trees is then chosen regarding the AUC measure.

I began the evaluation for the original training sample with a default rate of 2.6% and 26 covariates with 22 numerical and 4 categorical ones. By running 200 trees, the mtry value is changed and stepwise increased from 1 to 26 variables. Figure 6.7 shows the AUC values for the different mtry numbers trained on the original training sample and applied to the test sample. Note that for reasons of clarity, every fifth value is given for mtry 10 to 25. As shown in Figure 6.7, the highest predictive accuracy for 200 trees for the presented data can be achieved with the mtry value 6; i.e., that 6 variables are randomly selected for finding the best split.

In the next step, for tuning the parameters, the calculated mtry value of 6 is used to increase successively the number of trees. An extract of the results with the original training sample and the test sample is shown in Table 6.13. The best outcome is reached with 1000 trees and resulted in an AUC value of 0.6984.

As seen in Table 6.13, the prediction performance can still be improved by increasing the number of trees. However, the AUC value of the logit model with 0.7045 indicates that even with a high number of trees and optimized tuning parameters, the random forests

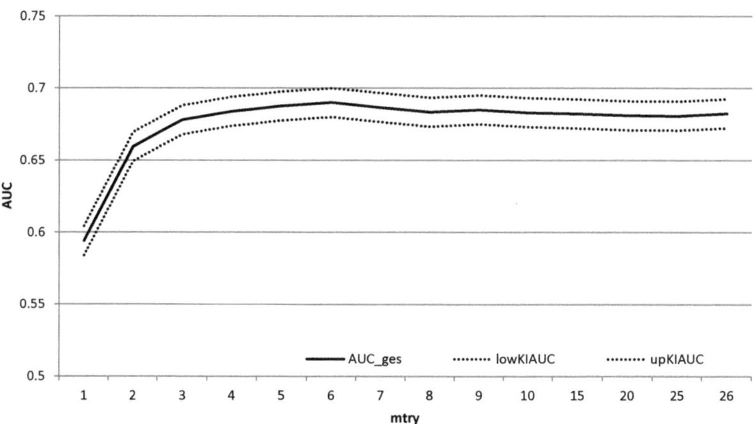

Figure 6.7: Illustration of the prediction accuracy for random forests trained on the original training sample with 2.6% default rate and applied to the test sample by varying the mtry-values for a fixed number of 200 trees.

underachieve compared to the logistic regression model on the original training sample. The validation data show that the predictive accuracy of the random forest algorithm trained on the original training data lies with 0.6964 under the AUC value of 0.7118 for the logit model (cf. Table 6.14). This outcome for the credit scoring data is surprising since random forests reach excellent results in other scientific fields and some credit scoring studies.

AUC	lowCI95%	upCI95%	ntree	mtry
0.5815	0.5714	0.5917	10	6
0.6477	0.6375	0.6578	50	6
0.6763	0.6662	0.6863	100	6
0.6873	0.6773	0.6973	200	6
0.6906	0.6806	0.7006	300	6
0.6941	0.6841	0.7041	400	6
0.6946	0.6846	0.7045	500	6
0.6984	0.6884	0.7083	1000	6

Table 6.13: Prediction accuracy of random forests with CART-like trees trained on the original training sample with continuous covariates and applied to the test sample with an mtry value of 6 and varying numbers of trees (ntree).

For further evaluation, the presented procedure for analyzing random forests and tuning the parameters is applied to the training samples with different default rates. Table 6.14 contains the final results for the test and validation samples by training on the original training data, as shown above, and also includes the results for the test sample trained on the training data with 20% default rate and 50% default rate.

		AUC	lowCI95%	upCI95%	ntree	mtry	LogReg	*p*-value
trainORG	test	0.6984	0.6884	0.7083	1000	6	0.7045	
	validation	0.6964	0.6826	0.7102	1000	6	0.7118	0.0671
train020	test	0.7202	0.7104	0.7299	2000	3	0.7054	
	validation	0.7184	0.7048	0.7320	2000	3	0.7129	0.4971
train050	test	0.7130	0.7031	0.7229	2000	2	0.7051	
	validation	0.7163	0.7027	0.7299	2000	2	0.7125	0.6373

Table 6.14: Random forest results with continuous variables trained on three different training samples, tuned on the test sample and applied to the validation sample in comparison to the logistic regression results. The *p*-values result from DeLong's test.

Among all mtry values from 1 to 26, the best mtry value of 3 was identified for the 20% training sample applied to the test sample. By fixing the mtry value to 3, the number of trees was varied for the mentioned training sample. The best outcome is reached for 2000 trees with an AUC value of 0.7202. Therefore, on the training sample with 20% default rate, the random forests outperform the logit model with an AUC value of 0.7054. To prevent overfitting and to validate the results, the outcomes are applied to the validation data also presented in Table 6.14. For the validation data, the tuning parameters are maintained. For the training on the 20% training sample the validation shows AUC values of 0.7184 for the random forest algorithm compared to 0.7129 for the logit model. In contrast to the results for the original training sample, the results on the 20% training sample show slightly better results for the random forest algorithm.

On the training data with 50% default rate and the test sample, the searching procedure for the optimal mtry value resulted in a value of 2. The best predictive accuracy is indicated with 2000 trees. The performance of the random forests for this credit scoring data lies above the performance of the logistic regression for the same data. The same result is demonstrated on the validation data (cf. Table 6.14). The AUC differences are not significant according to DeLong's test.

Different performance measures are included in Table 6.15 for the best random forest results with continuous covariates compared to the corresponding logit model results on the validation data.

		AUC	Gini	H measure	KS	MER	Brier score
trainORG	random forest	0.6964	0.3928	0.1089	0.2987	0.0272	0.0275
	LogReg	0.7118	0.4237	0.1283	0.3101	0.0272	0.0270
train020	random forest	0.7184	0.4369	0.1344	0.3232	0.0272	0.0682
	LogReg	0.7129	0.4257	0.1286	0.3143	0.0272	0.0719
train050	random forest	0.7163	0.4326	0.1290	0.3195	0.0272	0.2439
	LogReg	0.7125	0.4249	0.1276	0.3142	0.0272	0.2494

Table 6.15: Further measures for the random forest results with continuous variables compared to the logistic regression results for the validation data.

The different performance measures approve the previous findings. Since the logit model definitely outperforms the random forest for the validation of the original training sample, the random forest results are slightly better than the logit models for the validation of the 20% and 50% training samples. The H measure for the random forest trained on the 50% training sample and applied to the validation data is 0.1290 compared to 0.1276 for the logit model. The KS value also indicates a higher value. The MER is on the same level and the Brier score denotes better performance for the random forest algorithm.

The result for the original training sample is graphically demonstrated with the ROC-curve in Figure 6.8. The graph shows the false positive rate versus the true positive rate in order to create the ROC-curve for the validation data. This is done by using the original training data on which the algorithm was trained. Obviously, the black line for the random forest lies under the curve of the logistic regression (dashed line). This confirms the result for the original training data that the random forests can not improve the predictive performance here.

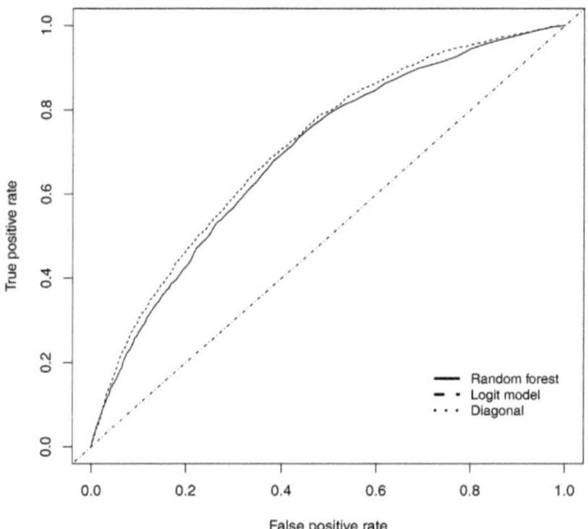

Figure 6.8: ROC graphs for the random forest (AUC 0.6964) and the logit model (AUC 0.7118) trained on the original training sample applied to the validation data.

An interesting aspect of choosing the tuning parameters is that relatively small values for mtry are selected. This is in line with the statements of Liaw and Wiener (2002) noting that even mtry values of 1 can give good performance for some data. While the performance decreases for higher mtry values, the predictive accuracy increases with higher ntree values. This certainly implies that the covariates more frequently have the chance to be chosen in the algorithm. By training the data with the three different training samples, all 26 variables are used within the random forest algorithm. By analyzing the frequencies of how often variables are used in the forest, it is shown that for all samples the covariates were used a hundred and even a thousand times within the forest. However, counting the variable frequencies in a forest does not take the discriminative power and the split position within a tree into account.

For this reason, the variable importance for the forests was examined with the importance measures mean decrease in node impurity and the mean decrease in accuracy. The former measures the reduction of node impurity by the Gini index, while the latter measures the prediction error on the out-of-bag data by permuting the predictor variable. Both importance measures are described in Section 6.3.2. Figure 6.9 contains two graphics for the random forest on the 20% training sample with the mean decrease gini in the upper one and the mean decrease accuracy in the lower one. Since all variables are used within the forest for the 20% sample, all covariates are shown in the figure. The higher the values of the measures are, the more important are the variables. High values of the importance measures consequently denote good predictor variables. The order of the covariates follows the height of the mean decrease gini values.

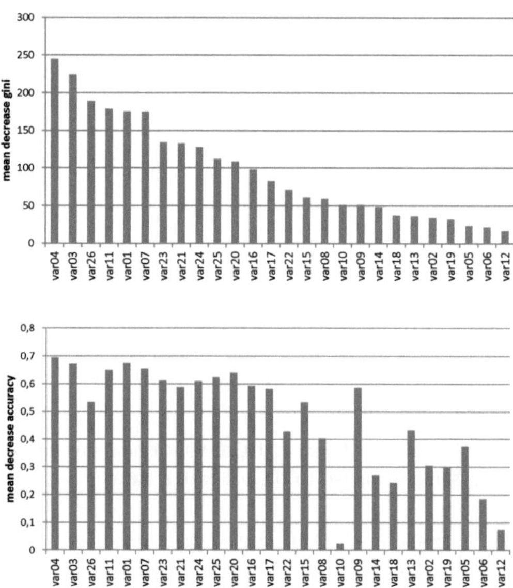

Figure 6.9: Mean decrease in node impurity and mean decrease in accuracy on the 20% training sample for the 26 covariates.

For the sample presented in Figure 6.9, both measures differ in several covariates from each other. But the most important variables are similar in both cases as well as the less important variables. The tendency for the importance of the covariates is equal for both measures. By comparing the mean decrease gini for the three different training samples, the variables show also similar importances. Since the samples with 20% and 50% default rate present a nearly identical variable order, the importance measure differs slightly on the original training sample. Figure A.5, with the three charts, is included in the Appendix. The first six variables are the most predictive ones even if the order differentiates for the original sample. The variables var05, var06 and var12 indicate low performance in all three cases.

In general, this information about the variables confirms the univariate findings from the logistic regression procedure described in Section 3.2. Figure 3.2 shows the univariate AUC values for the covariates. Variable var03 and var04 are the best univariate variables. Variable var05 and var06 are in the middle of the order even though these were the poorest in the variable importance measures. But note that this comparison only gives weak indications since the importance measures result from a multivariate process and the AUC values are univariate. Therefore, this comparison has to be seen with prudence.

As described in Section 6.1, the classical CART-algorithm favors numerical variables with many potential categories. The four categorical variables with only a few levels (var05, var06, var08, var09) indeed show lower importance in the random forest algorithms.

Before random forests are in the following tested in the framework of conditional inference trees, the best outcomes for the random forest results with coarse classification are presented in Table 6.16.

The best ntree values are 1000, 1000 and 1500 (trainORG, train020, train050) and the values for the tuning parameter mtry are 8, 3 and 3, respectively. Most of the performance measures show superior results for the logit model. The AUC differences are significant according to DeLong's test. For the 20% training sample, the validation results indicate a lower Brier score for the random forest algorithm. But all other measures show better performance for the logistic regression results.

		AUC	*p*-value	Gini	H measure	KS	MER	Brier score
trainORG	random forest	0.6075		0.2150	0.0408	0.1792	0.0259	0.0273
	LogReg	0.7263	0.0000	0.4527	0.1468	0.3310	0.0260	0.0260
train020	random forest	0.6994		0.3989	0.1156	0.2895	0.0260	0.0548
	LogReg	0.7282	0.0004	0.4564	0.1476	0.3314	0.0260	0.0658
train050	random forest	0.7093		0.4187	0.1210	0.3098	0.0260	0.2330
	LogReg	0.7275	0.0224	0.4550	0.1473	0.3364	0.0260	0.2217

Table 6.16: Different performance measures for the random forest results with classified variables compared to the logistic regression results for the validation data. The *p*-values result from DeLong's test.

The validation results for the 50% training sample also show superior results for the logit model. All different performance measures indicate better performance for the logistic regression. The presented random forest results, therefore, do not outperform the classical logit model. Only in a few cases, the random forest results indicate better performance compared to the logit model.

Random forests with conditional inference trees. The evaluation of random forests in credit scoring is continued with the conditional inference trees as basis for the random forest algorithm. Hereby, unbiased variable selection is guaranteed (cf. Section 6.1). The variable importances are examined with the permutation accuracy importance measure and the AUC-based importance measure.

The procedure for analyzing these random forests is similar to the procedure for random forests with CART-like trees. First, the tuning parameter mtry is chosen by running a fixed number of trees according to the highest AUC value. In a second step, the chosen mtry value is used by increasing the number of trees. The AUC measure is used as criterion for tuning the tuning parameters. Other performance measures are evaluated for the final results.

Computational reasons restrict the analysis to the 50%-training sample with 3.394 observations. Moreover, in one variable a few values had to be transformed in order to enable the prediction. This transformation has a negligible effect on the variable and training the data, but guarantees the inclusion of all 26 variables for the analysis. 22 covariates are numerical, while 4 are categorical. Data sets with encoded meaning were excluded from the samples in order to use continuous covariates. The algorithm is trained on the training sample, the tuning parameters are optimized concerning the AUC on the test sample, and the validation is examined on the validation sample.

The analysis is started by training the 50%-training sample for 200 trees, varying the mtry values from 1 to 26 and regarding the predictive accuracy on the test sample. The highest AUC value is reached with an mtry value of 7. For higher mtry values, the AUC measure is nearly on the same level and decreases for values greater than 16. Figure A.6 in the Appendix shows the curve with the different AUC values.

In the next step, the tuning parameter mtry is fixed to 7 while the number of trees was increased. Because of computational reasons, the ntree tuning parameter were computed with intervals of 50. The results trained on the 50%-training sample applied to the test sample are shown in Table 6.17.

The predictive accuracy increases slightly until 950 trees. The best random forest result with conditional inference trees is therefore 0.7197 on the test sample, trained on the 50%-training sample with 950 trees and a mtry value of 7. This predictive accuracy is higher than the original random forest algorithm, with an AUC value of 0.7130 on the test sample (cf. Table 6.14). The result is also higher than the logistic regression model with 0.7051 and continuous covariates. The corresponding AUC value of the logit model with coarse classification, however, denotes 0.7207.

In the presented credit scoring data there are covariates with different scales; i.e., numerical and categorical variables. The classical random forest algorithm favors numerical

AUC	lowCI95%	upCI95%	ntree
0.7042	0.6943	0.7141	10
0.7155	0.7057	0.7254	50
0.7174	0.7075	0.7272	100
0.7171	0.7072	0.7269	150
0.7181	0.7082	0.7279	200
0.7184	0.7086	0.7283	250
0.7189	0.7090	0.7287	300
0.7189	0.7090	0.7287	350
0.7187	0.7089	0.7285	400
0.7189	0.7091	0.7288	450
0.7192	0.7094	0.7290	500
0.7194	0.7095	0.7292	550
0.7195	0.7097	0.7293	600
0.7196	0.7097	0.7294	650
0.7195	0.7097	0.7293	700
0.7196	0.7098	0.7294	750
0.7195	0.7097	0.7293	800
0.7195	0.7097	0.7293	850
0.7195	0.7097	0.7294	900
0.7197	0.7098	0.7295	950
0.7197	0.7098	0.7295	1000

Table 6.17: Prediction accuracy of random forests with conditional inference trees trained on the 50% training sample with continuous covariates and applied to the test sample with a fixed mtry value of 7 and varying numbers of trees.

variables und covariates with many categories. The random forest algorithm with conditional inference trees, combined with subsampling, guarantees unbiased variable selection. However, even if the predictive accuracy on the credit scoring data can be improved with the advanced random forest algorithm being compared to the classical method, both random forest algorithms do not outperform the logit model with classified variables.

The results on the validation sample, shown in Table 6.18, confirm the previous findings.

		AUC	*p*-value	H measure	KS	MER	Brier score
test	random forest	0.7197		0.1389	0.3190	0.0275	0.2269
	LogReg	0.7051		0.1211	0.3065	0.0275	0.2380
validation	random forest	0.7241		0.1408	0.3264	0.0272	0.2317
	LogReg	0.7125	0.1528	0.1276	0.3142	0.0272	0.2494

Table 6.18: Prediction accuracy of random forests based on conditional inference trees trained on the 50% training sample with continuous covariates, tuned the parameters on the test sample and applied to the validation sample. The mtry value is 7, the ntree value denotes 950. The *p*-value results from DeLong's test.

The random forest based on conditional inference trees outperforms the original random forest with an AUC value of 0.7241 compared to 0.7163 (cf. Table 6.14). The result is also better than the logit model with an AUC value of 0.7125. The corresponding AUC value for the logit model with classified variables denotes 0.7224. For this case, the random forest algorithm outperforms the classical logit model.

An important aspect within random forest algorithms is the importance of the variables to show the relevance of the different covariates. The classical mean decrease in node impurity and the mean decrease in accuracy for the original random forest algorithm are already presented. Within the unbiased random forest algorithm, the permutation accuracy measure is analyzed in comparison to the original measure. The AUC-based variable importance measure described in Section 6.3.2 is also examined.

By analyzing the variable importance measures, it is important to check whether there are the same results for different random seeds before interpreting the ranking order (Strobl et al., 2009). If there are large differences in the tendency, the number of trees should be increased.

Figure 6.10 shows the three importance measures for all 26 variables for the 50% training sample. The figure at the top contains the permutation accuracy measure based on the random forest algorithm with unbiased variable selection, trained with 950 trees and a mtry value of 7. The figure in the middle indicates the AUC-based importance measure where the error rate is replaced by the AUC measure before and after permuting a covariate. For comparison reasons, the figure in the bottom shows the mean decrease accuracy from the original random forest algorithm. It is based on training the 50% training sample with 2000 trees and a value of 2 for the tuning parameter mtry. The order follows the ranking of the permutation accuracy measure. The results for different random seeds showed similar results and similar ranking orders.

The first two measures indicate similar results for the 26 predictor variables. Variables 3, 9 and 4 are indicated as the most relevant variables, while variables 12, 19 and 10 are least important. Only a few variables show a different ranking order, var07 and var25 for example, comparing these two measures. The general tendency is equivalent. Even if these results are quite similar, the AUC-based measure should be taken into account especially for unbalanced data (Janitza et al., 2013). In this analysis, the default rate denotes 50%, but the normal default rate in credit scoring is quite low as shown in the original training sample with 2.6%. For the credit scoring case presented here, random forests for the original data are computationally too intensive, but the AUC-based measure is an approach that should be generally considered by analyzing random forests for credit scoring.

The classical mean decrease in accuracy is also compared to the other variable importance measures and is shown in the bottom of Figure 6.10. Even if the most relevant variables (var03 and var04) and the variables with low relevance (var12, var19, var10) are ordered analogously, there are more differences in the ranking order. For instance, variable 9 is indicated as less important than in the other both measures. This can be explained with the unbiased variable selection of the newer random forest method because variable 9 denotes a categorical predictor variable. The other three categorical variables are variable 5, 6 and 8 and also show different ranking orders in the classical importance measure.

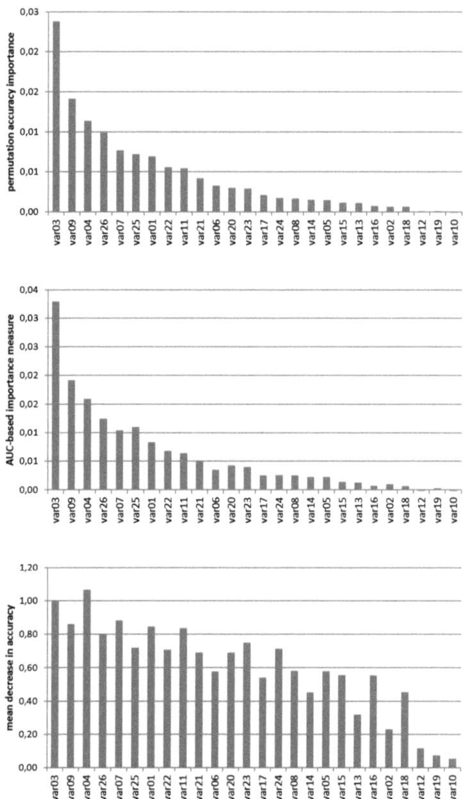

Figure 6.10: Variable importance measures on the 50% training sample: permutation accuracy importance (top), AUC-based variable importance (middle), and mean decrease in accuracy (bottom).

For confidentiality reasons, a more detailed analysis of the predictor variables concerning the contents is not possible. However, the differences imply to examine the various variable importance measures in random forest algorithms. Moreover, the conditional version of the importance measure described in Section 6.3.2 should be considered in a next step since there are correlated predictors in the credit scoring case. Actually, the amount of 26 variables and more than 3000 observations are computationally too intensive for the permutation scheme.

The evaluation of the random forests in credit scoring is completed by testing the algorithm with classified variables. The categories for the predictor variables are adopted from the classes of the logit model. The random forest with unbiased variable selection are used for the analysis. The same procedure is applied, as was previously done, for the before mentioned evaluations of random forests. While the three data sets (50% training, test and validation) are used for the examination, the AUC measure is used for choosing the tuning parameters.

The final results for the test and validation sample trained on the 50%-training sample are shown in Table 6.19 including different performance measures. The best value for the tuning parameter mtry is yielded with 6 randomly selected variables in each node. The number of trees is increased in steps of 50, due to computational reasons, and reached the best result with 700 trees.

		AUC	*p*-value	Gini	H measure	KS	MER	Brier score
test	random forest	0.7149		0.4297	0.1335	0.3061	0.0263	0.2169
	LogReg	0.7211		0.4423	0.1421	0.3255	0.0263	0.2121
validation	random forest	0.7193		0.4387	0.1346	0.3232	0.0260	0.2204
	LogReg	0.7275	0.3004	0.4550	0.1473	0.3364	0.0260	0.2217

Table 6.19: Best results of random forests based on conditional inference trees trained on the 50% training sample with categorized variables, tuned the parameters on the test sample and applied to the validation sample. The mtry value is 6, the ntree value denotes 700. The results are compared to the logistic regression outcomes. The *p*-value results from DeLong's test.

Compared to the logistic regression results, the predictive accuracy of the random forest on the test sample underachieves, with 0.7149 versus 0.7211. On the validation sample, the logit model yields a performance measured with the AUC of 0.7275 compared to 0.7193 of the random forest algorithm. This leads to the finding that in this evaluation scenario, the random forests can not outperform the predictive performance of the logit model. The AUC difference on the validation data is not significant according to DeLong's test.
The different performance measures approve the findings that the logit model outperforms the random forest in this case. Only the Brier score on the validation data indicates slightly better performance for the random forest algorithm, but all other measures definitely show better performance for the logit model.

For analyzing the relevance of the variables of the classified predictor variables, the permutation accuracy measure and the AUC-based importance measure are presented for the 26 classified predictor variables in Figure 6.11. It is interesting that variable 4 dominates the other covariates and shows the highest importance. Variable 4 was also very relevant in the former random forest algorithms, but variable 3 was often more important. The latter variable indicates less relevance in the classified version.

In general, it is remarkable that a lot of covariates display very low relevance, and the two measures show different ranking orders for the classified variables. This would imply to analyze the various variables with regard to their contents. However, this is not possible for the presented credit scoring case due to confidentiality reasons.

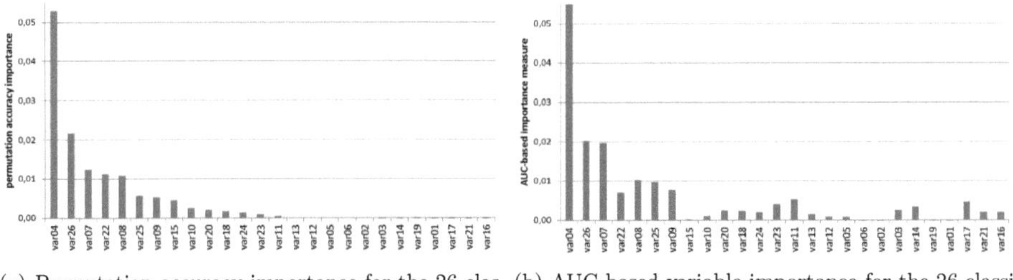

(a) Permutation accuracy importance for the 26 classified variables

(b) AUC-based variable importance for the 26 classified variables

Figure 6.11: Variable importance measures on the 50%-training sample with classified predictor variables.

In summary, it can be stated that the random forest algorithms obviously improve the predictive accuracy compared to single classification trees, and overcome the instability due to the aggregation of many trees. This improvement comes along with less interpretability of the models since single classification trees offer good interpretable rules. Also, for the presented credit scoring case, random forests can outperform the predictive performance of the logit model in some but not in all cases. The variable importance measures show interesting results for the predictor variables, and can therefore certainly make a contribution to understand, analyze and interpret the variables.

6.5 Discussion of Recursive Partitioning Methods for Credit Scoring

In this section new recursive partitioning algorithms are evaluated for the credit scoring case. The performance of the classification trees is rather poor compared to the predictive accuracy of the classical scorecard with logistic regression. The conditional inference trees presented here overcome the problem of unbiased variable selection and overfitting of the classical CART approach. Numerical covariates and variables with many potential cutpoints are not favored any more, and no form of pruning is needed for the new classification algorithm. However, the results still underachieve in the credit scoring case compared to the logit model. An advantage of classification trees is their binary structure and the good interpretability, which is still an important aspect in credit scoring. Since classification trees do not produce competitive results for the presented credit scoring case, from this point of view, they can not be recommended as an alternative to the classical scoring model.

The model-based recursive partitioning algorithm is an interesting approach for combining the classical logit model with the classification trees. The performance of the classical logit model can not be outperformed by the new model based approach, but the results are quite competitive. The stability of the variable coefficients is tested regarding the potential partitioning variables. This gives interesting insights in the variables and can help improving the logistic regression model even if the logistic regression based tree is not used for the final scoring model.

The presented random forest algorithms overcome the instability of single classification trees. In the current credit scoring evaluation, random forests perform better than single classification trees. In comparison with the benchmark results of the logit models, random forests outperform in some cases with respect to the AUC measure and other performance measures. But in some cases presented here, the logit model still yields better results compared to random forests. The random forests with conditional inference trees show better performance compared to the logit model. However, the results are different from the good performance of random forests in other scientific fields. And even other applications in the credit scoring area present higher predictive accuracies compared to the logit model (cf. for instance Lessmann et al. (2013)). Moreover, random forests are 'black box' models and hard to interpret by combining lots of single trees. Random forest algorithms present variable importance measures, which allow interesting investigations for analyzing the variables. These measures can provide added value to the univariate analysis in the classical scorecard development process.

The new methods are computationally very intensive and restrict the analysis in some parts to the 50%-training sample with around 3000 observations. Since large data sets are common for building scoring models in retail credit scoring, extensions for applying big data would be an important point for further developments.

Recursive partitioning algorithms are not yet implemented in many standard statistical software packages and therefore not easily applied in retail banks. But the development is going further with the implementation of not only classification trees but also other recursive partitioning techniques. This is an important step to develop these techniques in the credit scoring area and to benefit from the new capabilities.

Chapter 7

Boosting

7.1 Boosting Algorithms

Boosting methods are presented here as newer classification algorithms from machine learning. The basic idea for boosting is the aggregation of many weak classifiers in order to improve classification performance. AdaBoost (also commonly known as Discrete AdaBoost), introduced by Freund and Schapire (1996), is one of the most well-known algorithms. It aims to find the optimal classifier given the current distribution of weights in each step (Culp et al., 2006). Incorrectly classified observations with the current distribution of weights are weighted higher in the next iteration, while correctly classified observations are weighted lower. The final model outperforms with high probability any individual classifier concerning the misclassification error rate. Further important works on boosting follow with Breiman (1998) and Breiman (1999) demonstrating that the AdaBoost can be seen as an optimization algorithm and a gradient descent procedure in function space. The interpretation of boosting as additive stage-wise modeling from Friedman et al. (2000) describes another milestone for boosting in the statistical field. "Additive" in this context does not imply additivity in the covariates, but means boosting is an additive combination of weak learners (Bühlmann and Hothorn, 2007).

7.1.1 Rationale of Gradient Boosting

Friedman et al. (2000) propose different varieties of the AdaBoost algorithm like Real, Logit and Gentle AdaBoost. For illustrating the general algorithm, I followed the description of Culp et al. (2006) based on Friedman (2001) in Table 7.1. The aim is to find a classification rule $F(x)$ for prediction accuracy. Boosting should find the optimal classifier by minimizing an underlying loss function using a gradient descent search to define the weights and the learner in each step.

The general algorithm presents an iterative method for optimizing an arbitrary loss function. Culp et al. (2006) propose the exponential ($L(y, f) = e^{-yf}$) and logistic ($L(y, f) = log(1+e^{-yf})$) loss functions. The regularization parameter ν denotes the learning rate. The special functions of η define the type of boosting algorithm: $\eta(x) = sign(x)$ for Discrete

---

Initialize $F(x) := 0$

for $m = 1$ to M **do**

Set $w_i = -\frac{\partial L(y,g)}{\partial g}|_{g=F(x)}$

Fit $y = \eta(h_m(x))$ as the base weighted classifier using $|w_i|$, with training sample π_m

Compute line search step $\alpha = arg\ min_\alpha \sum_{i\in\pi_m} L(y_i, F(x) + \alpha\eta(h_m(x_i)))$

Update $F(x) = F(x) + \nu\alpha_m\eta(h_m(x))$

end for

Table 7.1: Stochastic Gradient Boosting Algorithm.

AdaBoost, $\eta(x) = 0.5log(\frac{x}{1-x})$ for Real AdaBoost and $\eta(x) = x$ for Gentle AdaBoost. The definition for predicting the probability class estimates for the different boosting algorithms is $\hat{P}(Y = 1 \mid x) = \frac{e^{2F(x)}}{1+e^{2F(x)}}$.

7.1.2 Component-Wise Gradient Boosting

Based on the former boosting algorithms, Bühlmann and Yu (2003) present the component-wise gradient boosting in their paper. The fact that the covariates are fitted separately against the gradient is the main advancement compared to existing algorithms. Another advantage of the method is it allows variable selection during the fitting process.

The basic idea is still to model the outcome variable y given a set of covariates x_i and finding the optimal prediction. In the component-wise gradient boosting, this is realized by minimizing a loss function $\rho(y, f) \in \mathbb{R}$ over the prediction f. The descriptions follow Bühlmann and Hothorn (2007) and Hofner et al. (2014). The aim is to estimate the "optimal" prediction function f^* defined as

$$f^* := \arg\min_f \mathbb{E}_{Y,X}[\rho(Y, f(X))] \tag{7.1}$$

where the loss function $\rho(Y, f)$ is assumed to be differentiable with respect to f. Since in practice the theoretical mean given in 7.1 is usually unknown, boosting algorithms minimize the empirical risk $\mathcal{R} := \sum_{i=1}^{n} \rho(y_i, f(x_i))$ over f. The empirical risk $\mathcal{R}$ is minimized over f in the following way:

1. Initialize an n-dimensional vector $\hat{f}^{[0]}$ with offset values.

2. Specify the base-learners for the covariates. Set $m = 0$.

3. Increase m by 1. Compute the negative gradient $-\frac{\partial \rho}{\partial f}$ of the loss function. Evaluate the negative gradient at the estimate of the previous iteration $\hat{f}^{[m-1]}(x_i)$, $i = 1, \dots, n$. This yields the negative gradient vector

$$u^{[m]} = \left(u_i^{[m]}\right)_{i=1,\dots,n} := \left(-\frac{\partial}{\partial f}\rho\left(y_i, \hat{f}^{[m-1]}(x_i)\right)\right) \tag{7.2}$$

4. Fit the negative gradient vector $u^{[m]}$ separately to each of the P covariates using the base-learners defined in step 2. This results in P vectors of predicted values, where each vector denotes an estimate of the negative gradient vector $u^{[m]}$.

5. Regarding the residual sum of square (RSS) criterion, select the base-learner with the best fitting of $u^{[m]}$. Set $\hat{u}^{[m]}$ equal to the fitted values of the best base-learner.

6. Update $\hat{f}^{[m]} = \hat{f}^{[m-1]} + \nu\, \hat{u}^{[m]}$, with $0 < \nu < 1$ as a step length factor.

7. Iterate steps 3 to 6 until $m = m_{stop}$ for a given stopping iteration m_{stop}.

Further explanations to the base-learners and the tuning parameters step length ν and iterations m_{stop} follow below. The above algorithm shows the descending along the gradient of the empirical risk $\mathcal{R}$. In each iteration, an estimate of the true negative gradient of $\mathcal{R}$ is added to the estimate of f and a structural relationship between y and the selected covariates is established (Hofner et al., 2014). Since only one base-learner is used for the update of $\hat{f}^{[m]}$, step 5 and 6 denote the variable selection of the algorithm. This variable choice implies that a base-learner can be chosen several times and that the estimates also can be equal to zero, when the base-learner might not have been selected. In the former case, the function estimate is defined by the sum of the individual estimates $\nu \cdot \hat{u}^{[m-1]}$ from the iterations where the base-learner was chosen.

By specifying the appropriate loss function $\rho(\cdot,\cdot)$, a huge variety of different boosting algorithms arise. An extensive overview of loss functions and boosting algorithms is given in Bühlmann and Hothorn (2007). Based on the credit scoring context, the focus is on binary classification with a response variable $Y \in \{0,1\}$ and $\mathbb{P}[Y = 1 \mid x] = \pi(f)$. The exponential loss, leading to the famous AdaBoost algorithm, is previously mentioned in Section 7.1.1. In addition, the negative binomial log-likelihood (Bühlmann and Hothorn, 2007) is considered

$$\rho(y,f) = -\left[y\log(\pi(f)) + (1-y)\log(1-\pi(f))\right] \tag{7.3}$$
$$= \log(1+\exp(-2\tilde{y}f)) \tag{7.4}$$

where $\tilde{y} = 2y - 1$. The transformation of the binary response with $\tilde{y} \in \{-1, 1\}$ results due to computational efficiency (Hofner et al., 2014). The presented loss function, offers the opportunity to fit a logistic regression model via gradient boosting. In addition, the AUC measure can be used as loss function to optimize the area under the ROC curve. This topic is covered in detail in Section 7.3.

The stopping iteration m_{stop} is the most important tuning parameter in boosting algorithms. At the beginning, a debate existed as to whether boosting algorithms are resistant against overfitting. In the meantime, Bühlmann and Hothorn (2007) state that an early stopping (before running to convergence) is necessary and refer to Jiang (2004) and Bartlett and Traskin (2007). Therefore, in component-wise gradient boosting, cross-validation techniques and AIC-based techniques are proposed to choose the optimal stopping iteration (Bühlmann, 2006). Since the AIC-based procedure tends to exceed the optimal iteration number (Mayr et al., 2012), it is recommended to use cross-validated estimates of the empirical risk to determine an optimal number of boosting iterations. Note, however, that the AdaBoost is quite resistant against overfitting by increasing the iteration number and shows a slow overfitting behaviour (Bühlmann and Hothorn, 2007).

The step length factor ν is a less important tuning parameter for the performance of the boosting algorithm (Bühlmann and Hothorn, 2007). According to Bühlmann and Hothorn (2007), it is only required to choose a small value for ν (e.g. $\nu = 0.1$). With a smaller value of ν an increasing number of iterations is required for the boosting algorithm, while the influence on the predictive accuracy is rather low.

7.1.3 Base Learners

For the different types of boosting and the structural assumption of the model, the specification of base-learners is required. Bühlmann and Hothorn (2007) give an overview of different base procedures. Linear and categorical effects can be fitted in the sense of ordinary least squares. By using the negative binomial log-likelihood as loss function combined with linear least squares base-learner, the fit of a linear logistic regression model via gradient boosting arise (Bühlmann and Hothorn, 2007). For adding more flexibility to the linear structure of generalized linear models, smooth effects can be specified using P-splines (cf. Schmid and Hothorn, 2008). Hereby, a logistic additive model fit is obtained by using component-wise smoothing splines with negative binomial log-likelihood loss. Additionally, tree-based base learners are quite popular in the machine learning community. This base procedure is not fitted in the sense of penalized least squares (Hofner et al., 2014). Section 6.1 is referenced for detailed descriptions on classification and regression trees. Further explanations on base procedures, like bivariate P-Splines or random effects, can be found in Bühlmann and Hothorn (2007) and Hofner et al. (2014). By the combination of loss functions and base procedures, an extensive amount of possibilities for various boosting algorithms arise. In alignment with this thesis and the classification problem in credit scoring, the focus is on the base procedures in boosting algorithms on tree-based base learner, logistic regression and generalized additive models.

7.2 Boosting for Credit Scoring

Newer machine learning algorithms like boosting, especially component-wise gradient boosting, are less examined in the credit scoring literature. Remarks and references are already given in Section 6.4 with emphasis on recursive partitioning. At this point, some related work to boosting in credit scoring is outlined. For instance, the paper of Brown and Mues (2012) is mentioned where superior results for gradient boosting are indicated for the credit scoring case compared to other classifiers. Wang et al. (2011) propose a performance comparison for bagging and boosting with exceeding results compared to standard models. Finlay (2011) propose an error trimmed boosting algorithm which "successively removes well classified cases from the training set". In their study, which is based on an application and a behavioral data set, the results of this boosting algorithm outperform other classifiers including the AdaBoost.

In the credit scoring evaluation, the classical gradient boosting based on Friedman (2001) is analyzed. Moreover, it is concentrated on the boosting of logistic regression and generalized additive models within the component-wise gradient framework. Boosting with the AUC as a loss function is covered in Section 7.3.1.

7.2.1 Discrete, Real and Gentle AdaBoost

To begin the evaluation of the boosting algorithms in credit scoring, the Discrete, Real and Gentle AdaBoost were examined with tree base learners. In addition to the number of iterations, two tuning parameters are varied. The maxdepth parameter defines the forefathers one node can have. A maxdepth value of 1 is a so-called stump and implies that the node is only divided once. An increasing maxdepth value represents more complicated trees. The minsplit parameter describes the minimal number of observations required in a node to continue dividing that particular node further.

For training the data, the established training samples are used with the various default rates. Optimizing the tuning parameters required the test sample, while the validation that followed is conducted with the validation sample. The iterations are increased from 10 to 150, the maxdepth values from 1 to 30, and the minsplit values are tested with 0, 10 and 100. The same procedure is applied for Discrete, Real, and Gentle AdaBoost with exponential loss generating plenty of combinations. The results shown here represent the summary of the best outcomes concerning the AUC measure.

Table 7.2 shows the AUC values for the test and validation samples with the best tuning parameters in comparison to the logit model. Continuous variables are used for these results and the p-values are calculated using DeLong's test.

Trained on the original sample with 2.6% of defaults, 150 iterations, a maxdepth value of 30, and a minsplit value of 0, an AUC of 0.7256 is calculated for the test sample. The corresponding value from the logit model (0.7045) is significantly different. The analyses on the other samples show superior results on the test and validation samples, even though DeLong's test shows that the differences are not significant.

		AUC	iterations	maxdepth	minsplit	AUC LogReg	p-value
trainORG	test	0.7256	150	30	0	0.7045	
	validation	0.7258	150	30	0	0.7118	0.0899
train020	test	0.7240	150	4	0	0.7054	
	validation	0.7273	150	4	0	0.7129	0.0781
train050	test	0.7263	150	5	0	0.7051	
	validation	0.7249	150	5	0	0.7125	0.1257

Table 7.2: Gentle AdaBoost results with continuous variables trained on three different training samples and applied to the test and validation samples in comparison to the logistic regression results. The p-values result from DeLong's test.

Figure 7.1 shows the ROC graphs for the validation sample trained on the 20% training sample with continuous covariates. Although the distance in the graph is slight, the ROC generated by the logit model lies below that of the Gentle AdaBoost model.

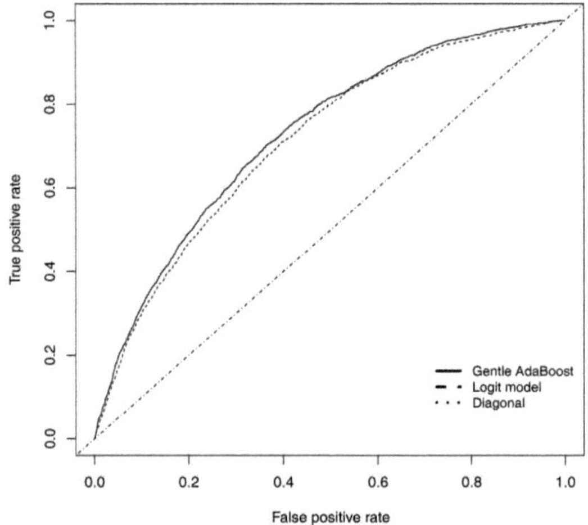

Figure 7.1: ROC graphs for the Gentle AdaBoost (AUC 0.7273) and the logit model (AUC 0.7129) trained on the 20% training sample applied to the validation data with continuous covariates.

Table 7.3 contains different performance measures comparing the Gentle AdaBoost with the logit model using the validation data and continuous variables. While the H measure and the KS indicate superior results for the boosting method over all samples, the MER is at the same level, and the Brier score shows better prediction performance for the 50% training sample.

		AUC	Gini	H measure	KS	MER	Brier score
trainORG	Gentle	0.7258	0.4515	0.1433	0.3372	0.0272	0.0270
	LogReg	0.7118	0.4237	0.1283	0.3101	0.0272	0.0270
train020	Gentle	0.7273	0.4546	0.1473	0.3381	0.0272	0.0715
	LogReg	0.7129	0.4257	0.1286	0.3143	0.0272	0.0719
train050	Gentle	0.7249	0.4498	0.1402	0.3330	0.0271	0.3322
	LogReg	0.7125	0.4249	0.1276	0.3142	0.0272	0.2494

Table 7.3: Further measures for the Gentle AdaBoost results with continuous variables compared to the logistic regression results for the validation data.

Similar conclusions can be shown for the comparison of the Gentle AdaBoost and the logistic regression with classified variables. Table 7.4 contains the outcomes for the validation data and different performance measures. The differences of the AUC measures are not significant according to DeLong's test, but the tendency of the other measures also indicates superior results for the boosting method. However, the Brier score only indicates better predictive performance for the 50% training sample.

		AUC	Gini	H measure	KS	MER	Brier score
trainORG	Gentle	0.7266	0.4532	0.1451	0.3377	0.0260	0.0259
	LogReg	0.7263	0.4527	0.1468	0.3310	0.0260	0.0260
train020	Gentle	0.7296	0.4593	0.1495	0.3373	0.0259	0.0557
	LogReg	0.7282	0.4564	0.1476	0.3314	0.0260	0.0658
train050	Gentle	0.7315	0.4630	0.1470	0.3417	0.0260	0.3081
	LogReg	0.7275	0.4550	0.1473	0.3364	0.0260	0.2217

Table 7.4: Further measures for the Gentle AdaBoost results with classified variables compared to the logistic regression results for the validation data.

The results for the Real and Discrete AdaBoost with continuous covariates are summarized in the Appendix (Tables A.6 to A.9). Since the AUC values and the other performance measures show superiority for the Real AdaBoost, the Brier score indicates better results for the logit model. The Discrete AdaBoost results outperform the corresponding logit model outcomes in terms of all performance measures. Not all AUC differences, however, are significant concerning DeLong's test.

7.2.2 Boosting Logistic Regression

Within the scope of boosting methods, the logistic regression models are estimated via gradient boosting. The negative binomial log-likelihood loss is used as a loss function. Due to computational reasons, the 50% training sample is used and the results are applied to the test and validation sample. The evaluation was started by using 26 explanatory variables (22 continuous, 4 categorical) and excluding data sets with encoded meanings. The comparable AUC values for the logit model are for the test sample 0.7051 with continuous and 0.7207 with classified variables. For the validation sample, the predictive power with continuous variables is 0.7125 and 0.7224 with classified variables.

Table 7.5 contains the results for the test and validation sample for different iteration numbers. This shows that the boosting results exceed the classical logistic regression model with continuous variables, but not the predictive performance of the models with classified covariates. The test and validation samples indicate the same conclusion. This finding is still true with a high number of iterations of 1000 or 1500 iterations. Moreover, the number of iterations is the most important tuning parameter in this case.

test			validation			
AUC	lowCI95%	upCI95%	AUC	lowCI95%	upCI95%	iterations
0.6779	0.6678	0.6879	0.6841	0.6703	0.6980	10
0.6995	0.6895	0.7094	0.7052	0.6915	0.7189	50
0.7052	0.6953	0.7151	0.7112	0.6975	0.7248	100
0.7078	0.6979	0.7177	0.7140	0.7003	0.7276	150
0.7103	0.7004	0.7202	0.7172	0.7036	0.7308	270
0.7110	0.7011	0.7208	0.7182	0.7046	0.7318	331
0.7113	0.7014	0.7212	0.7187	0.7051	0.7323	379
0.7118	0.7019	0.7217	0.7194	0.7058	0.7330	500
0.7127	0.7028	0.7225	0.7207	0.7071	0.7343	1000
0.7127	0.7028	0.7225	0.7206	0.7071	0.7342	1500

Table 7.5: Prediction accuracy for boosting logistic regression trained on the 50% training sample, applied to the test and validation samples with continuous covariates.

To prevent overfitting, it is important to determine the number of boosting iterations. As mentioned above (Section 7.1.2), the AIC is not recommended to find the optimal model. Therefore, cross-validated estimates of the empirical risk are used to extract the iteration number. The k-fold cross validation is estimated with 5 and 10 replicates.

Figure 7.2 shows two graphics with the number of iterations on the x-axis and the negative binomial likelihood on the y-axis. The appropriate number of boosting iterations is in the first case 379, and in the latter 331. The exact results with the AUC values are also contained in Table 7.5. In addition, the predictive risk for the 25 bootstrap samples is displayed and a number of 270 iterations is received (cf. Figure A.7 in the Appendix). As shown in Table 7.5, the improvements for the results with a higher number of iterations are slight.

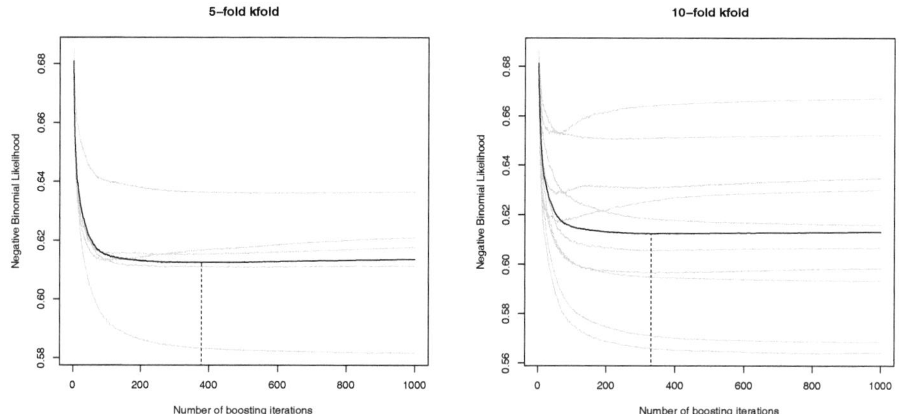

(a) 5-fold cross validation - appropriate number of boosting iterations 379

(b) 10-fold cross validation - appropriate number of boosting iterations 331

Figure 7.2: Results of the k-fold cross validation to extract the appropriate number of boosting iterations for boosting logistic regression.

Different performance measures are shown in Table 7.6 for the boosted logit model with 331 iterations and continuous covariates. For the validation data, the difference of the AUC measures is not significant according to DeLong's test. The other performance measures, however, indicate higher predictive performance for the boosted logit model. The MER is on the same level. These results confirm the previous findings regarding the AUC measure.

		AUC	*p*-value	H measure	KS	MER
test	Boost LogReg	0.7110		0.1295	0.3078	0.0275
	LogReg	0.7051		0.1211	0.3065	0.0275
validation	Boost LogReg	0.7182		0.1359	0.3193	0.0272
	LogReg	0.7125	0.4884	0.1276	0.3142	0.0272

Table 7.6: Different performance measures for boosting logistic regression with 331 iterations and continuous variables compared to the logistic regression results for the test and validation samples. The *p*-value results from DeLong's test.

A very important aspect for the early stopping is the complexity of the model. The model built with 1000 iterations includes 24 of the 26 available predictors. While the models with 270, 331 and 379 iterations contain 18 explanatory variables, these are the same for all three iteration numbers. The variables selected for the logistic model are included in the 18 predictors. The number of iteration reduces the complexity of the model and chooses more influential variables.

Even though, models are tested with the 8 variables from the logit model. This leads to the same AUC values as for the classical logistic regression (0.7051 on the test sample and 0.7125 on the validation sample with 1000 iterations).

In addition to the cross validation, Figure 7.3 displays the AUC values and the confidence intervals for the boosting iterations from 1 to 1000 for the test sample. The analogue graphic for the validation sample is in the Appendix (Figure A.8). Even if the charts show values up to 1000 iterations, the improvement of the predictive power is low after a few hundreds of iterations. After approximately 400 iterations, the curve stagnates. This corresponds with the appropriate iterations from cross validation.

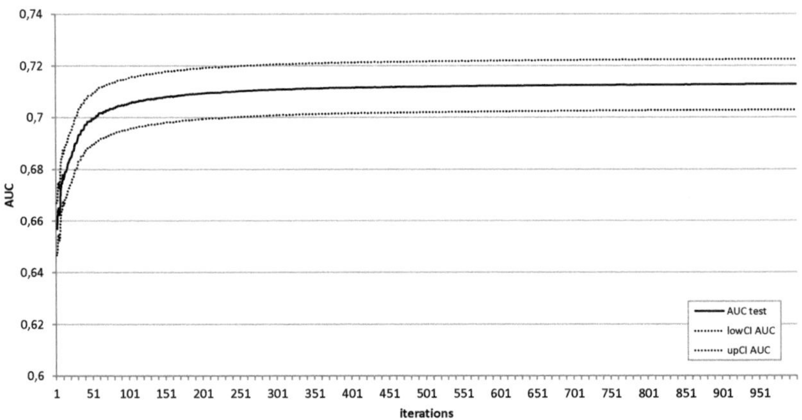

Figure 7.3: AUC values with confidence intervals for boosting logistic regression on the test sample for iterations 1 to 1000 with 26 potential continuous covariates.

While analysing boosting logistic regression models, categorical variables are also tested as used in the classical logistic model. I used the categories evaluated from the classical scorecard development. The proceeding is analogous to the previous description for boosting logistic regression, with the difference to use the 26 predictors as categorical effects. The results are shown in Table A.10 in the Appendix. Two charts with the outcomes of the iterations 1 to 1000 are also included in the Appendix (Figure A.9 for the test sample and Figure A.10 for the validation sample). The 50% training sample was used without excluding data sets. The appropriate number of iterations determined with cross validation (5- and 10- fold kfold, 25-fold bootstrap) rose high iteration numbers with 1499, 1498 and 1447. For boosting the model with 1500 iterations, the AUC value is 0.7273 (LogReg 0.7210) for the test sample and 0.7335 (LogReg 0.7225) for the validation sample. The values in brackets indicate the results for the classical logistic regression models. The high number of iterations implies quite complex models.

Table 7.7 contains the AUC measures for the boosted logit model with 1500 iterations and coarse classification compared to the classical logit model. Further performance measures are also included in the table.

		AUC	p-value	H measure	KS	MER
test	Boost LogReg	0.7273		0.1489	0.3297	0.0263
	LogReg	0.7210		0.1408	0.3257	0.0275
validation	Boost LogReg	0.7335		0.1537	0.3441	0.0260
	LogReg	0.7225	0.1645	0.1411	0.3305	0.0272

Table 7.7: Different performance measures for boosting logistic regression with 1500 iterations and classified variables compared to the logistic regression results for the test and validation samples. The p-value results from DeLong's test.

The H measure, KS and MER show better performance for the boosted logit model compared to the classical logistic regression. DeLong's test, however, indicates that the difference of the AUC measures on the validation data is not significant. Since boosting yields better results for 1500 iterations, complex models are evaluated in this case.

7.2.3 Boosting Generalized Additive Model

To add more flexibility to the linear structure in generalized linear models, logistic additive models are estimated via component-wise gradient boosting. P-Spline base-learners for smooth effects are used for 22 explanatory variables, while 4 variables are included as categorical effects. Equal to boosting logistic regression, the negative binomial log-likelihood loss is used to estimate the empirical risk. Data sets with encoded meaning are excluded from the 50% training sample that is used to train the boosting algorithm. The default ($\nu = 0.1$) is used for the step length factor. The results are applied to the test and validation sample. As previously mentioned above (Section 7.2.2), the predictive performance from the logit model on the test sample is 0.7207 with classified variables (0.7051 with continuous covariates) and on the validation sample 0.7224 (0.7125). Since iterations from 1 to 1000 were run, an extract of the outcomes is given in Table 7.8.

The results for the test sample, as well as for the validation sample, increase with the number of boosting iterations. Since the iteration number is the most important tuning parameter within component-wise gradient boosting, cross-validated estimates of the empirical risk are used to determine the appropriate number of boosting iterations. Figure 7.4 shows the estimated k-fold cross validation with 5 and 10 replicates. The plots display the empirical risk on the cross validation samples for iterations from 1 to 1000. The optimal number of boosting iterations results in the first case in 243, and in the latter case in 206 boosting iterations. The result for the 25-fold bootstrap samples is contained in the

test			validation			
AUC	lowCI95%	upCI95%	AUC	lowCI95%	upCI95%	iterations
0.6766	0.6666	0.6867	0.6839	0.6700	0.6978	10
0.7108	0.7009	0.7207	0.7178	0.7042	0.7314	50
0.7179	0.7081	0.7278	0.7247	0.7112	0.7383	100
0.7211	0.7118	0.7314	0.7265	0.7151	0.7422	150
0.7212	0.7114	0.7310	0.7266	0.7131	0.7402	203
0.7212	0.7114	0.7310	0.7266	0.7131	0.7402	206
0.7217	0.7119	0.7315	0.7267	0.7131	0.7402	243
0.7273	0.7176	0.7371	0.7325	0.7190	0.7459	500
0.7282	0.7184	0.7379	0.7314	0.7179	0.7449	1000

Table 7.8: Boosting generalized additive models with P-spline base-learners and the negative binomial log-likelihood loss trained on the 50% training sample, applied to the test and validation samples.

Appendix (Figure A.11). The optimal stopping iteration of 203 minimizes the average risk over all 25 samples.

The AUC values for the predictive performance of the models at the optimal stopping iterations are also presented in Table 7.8. As demonstrated in this table, the boosted logistic additive models outperform the results from the classical logit model.

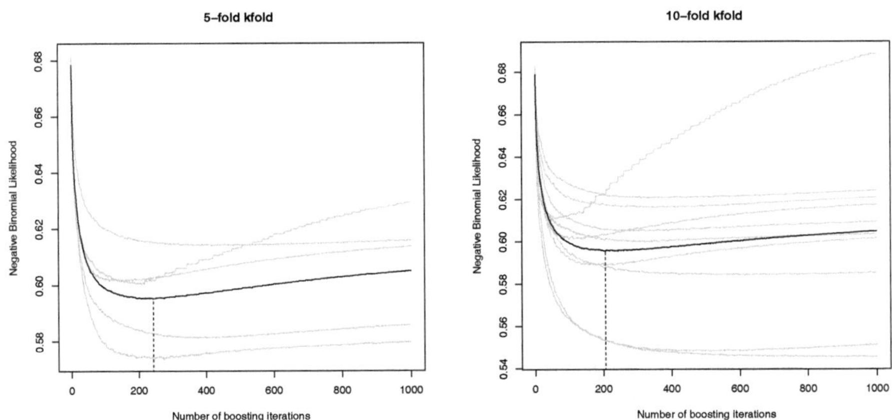

(a) 5-fold cross validation - appropriate number of boosting iterations 243

(b) 10-fold cross validation - appropriate number of boosting iterations 206

Figure 7.4: k-fold cross validation to extract the appropriate number of boosting iterations for boosting generalized additive models.

On the test sample, the boosted model with 243 iterations indicates an AUC value of 0.7217 compared to the predictive accuracy of 0.7207 from the logit model with classified variables (0.7051 with continuous covariates). The performance comparison on the validation sample indicates the same conclusion. AUC values around 0.7266 exceed the predictive performance of 0.7224 from the logistic regression with coarse classification (0.7125 with continuous variables).

Table 7.9 shows different performance measures for the results of the boosted additive model compared to the classical logistic regression. The boosted model was evaluated with 243 iterations. The H measures and the KS values indicate better performance for the boosted model. The results for the MER are on the same level. The presented measures in this table approve the previous findings.

		AUC	p-value	H measure	KS	MER
test	Boost GAM	0.7217		0.1413	0.3288	0.0275
	LogReg	0.7051		0.1211	0.3065	0.0275
validation	Boost GAM	0.7267		0.1438	0.3352	0.0272
	LogReg	0.7125	0.0815	0.1276	0.3142	0.0272

Table 7.9: Different performance measures for boosting the logistic additive model with 243 iterations compared to the logistic regression results with continuous variables for the test and validation samples. The p-value results from DeLong's test.

As described in Section 7.1.2, the component-wise gradient boosting includes variable selection during the optimization process. Therefore, the stopping iteration number also influences the complexity of the model. The model with 243 iterations involves 22 variables, while the model with 100 iterations includes 16 covariates.

This relatively high number of included base-learners follows to the test, where only the eight explanatory variables from the logit model are used as potential covariates. The procedure is the same as described with 26 variables. The training was on the 50%-sample, the application was to the test and validation sample and the choice of the optimal stopping iteration was defined via cross validation. An excerpt of all iteration numbers from 1 to 1000 is shown in Table A.11 in the Appendix. For the optimal iteration number of 261, all eight variables are included in the model with an AUC value on the test sample of 0.7219 (0.7207 logit model). The related validation results are 0.7266 to 0.7224 for the logit model. This result indicates that estimating the generalized additive model via gradient boosting still slightly outperforms the classical logistic regression, when only eight variables are used as potential covariates.

In Chapter 5, the additive logistic regression model is explained and evaluated for the credit scoring case. The models also include the eight explanatory variables of the classical logit model. Therefore, the results for the boosted GAM models with 8 variables are compared to the additive logistic models from Section 5.2. The related AUC values are taken from Table 5.1 from the mentioned chapter. On the test data, the classical GAM yields an AUC value of 0.7207. The corresponding AUC value of the boosted GAM denotes

0.7219. The predictive accuracy for the validation sample is 0.7263 for the additive logit model, compared to 0.7266 for the boosted additive logistic model. This implies that the boosting algorithm slightly outperforms the classical additive logit model. The flexibility of the model and the inclusion of nonlinearities in combination with boosting the model, shows good performance compared to the classical models.

The improvement in predictive accuracy is obviously obtained from the smoothness of the base-learners. Figure 7.5 displays the partial effects of three explanatory variables estimated as P-spline base-learners in the boosted generalized additive model with 243 iterations.

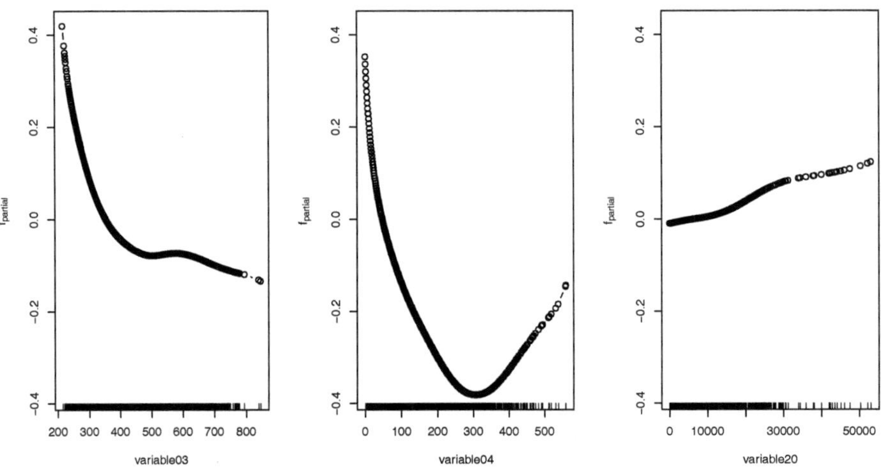

Figure 7.5: Partial effects for three explanatory variables estimated as P-spline base-learners in a boosted generalized additive model with 243 iterations.

As shown in Figure 7.5, variable03 and variable04 obviously have a non-linear relationship to the default information in the data. Variable20 seems to have a relationship relatively close to a linear function where most of the observations lie. However, especially with the higher values of the predictor, the relationship seems to be non-linear.

7.3 Boosting and the Optimization concerning AUC

In the research literature, different approaches are proposed for optimizing the AUC in the machine learning community and related areas. A short overview is given in the following.

Yan et al. (2003) consider an approximation of the Wilcoxon statistic to optimize the AUC directly with gradient based methods. They apply their algorithm to two data sets for customer behavior prediction. Cortes and Mohri (2003) discuss the relationship between the AUC and the error rate and state that algorithms with the aim to minimize the error rate "may not lead to the best possible AUC values". They indicate the connection between RankBoost (Freund et al., 2004) and the AUC optimization measured with the Wilcoxon statistic and present superior results in comparison to the former mentioned algorithm of Yan et al. (2003). However, they state that comparing RankBoost to AdaBoost shows similar results, which is a task for further analysis. Herschtal and Raskutti (2004) propose a direct optimization of the AUC within a gradient descent search as a linear binary classifier. Their empirical study is based on eight data sets from different fields. They show outperforming outcomes of the AUC optimization compared to other classifiers like linear regression and support vector machines. The work of Ma and Huang (2005) for biomarker selection with microarray data is previously mentioned in Chapter 4. Long and Servedio (2007) show the effects of boosting the AUC based on AdaBoost with the analysis of the presence of independent missclassification noise, but no empirical results are given. In another paper, Calders and Jaroszewicz (2007) focus on the efficiency of optimizing the AUC within gradient descent methods using a polynomial approximation of the AUC. The running time is comparable to linear discriminant analysis, which is below the computation time of the mentioned support vector machines. The AdaBoost algorithm is also extended by Sheng and Tada (2011) who introduce a weighted AUC formula and yield good results with their proposed AUCBoost method compared to AdaBoost with C4.5 classification trees as base learner. Another boosting method for maximizing an approximated smooth function of the AUC for low-dimensional data is proposed by Komori (2011). Komori and Eguchi (2010) extend the former work to high-dimensional settings by maximizing the pAUC. In addition, Wang (2011) propose the so-called HingeBoost for optimizing the hinge loss to maximize the AUC approximately.

The applications mainly concentrate on scientific fields like biostatistics, bioinformatics or medicine. Most of the studies compare the empirical results to other AUC based methods, rather than to the logistic regression model. The evaluation in the financial field, and especially in the consumer credit scoring area, is remarkably rare.

7.3.1 AUC as a Loss Function in Boosting Algorithms

The idea to use the AUC as direct optimization criterion is introduced in the boosting framework for the presented credit scoring case.

For the analysis, to compare predictive accuracy in the consumer credit market, the AUC is used as a loss function within the component-wise gradient boosting proposed by Bühlmann and Hothorn (2007). Instead of calculating with logistic or exponential loss, the

AUC measure is introduced as a loss function, which is implemented as $1 - AUC(y, f)$ with the definition (Hothorn et al., 2012)

$$AUC = (n_{-1}n_1)^{-1} \sum\nolimits_{i:y_i=1} \sum\nolimits_{j:y_j=-1} I(f_i > f_j) \tag{7.5}$$

As mentioned in Chapter 4, this is not differentiable. In this case, the distribution function of the triangular distribution on $[-1, 1]$ with mean 0 is used as an approximation of the jump function $I((f_i - f_j) > 0)$ (Hothorn et al., 2012).

The mentioned implementation of the AUC contains a division by the standard deviation $sd(f)$. This standardization in each step leads to different values for ν with different variances. Moreover, there is no σ implemented in the function to control the approximation to the jump function. For that reason,the PAUC function proposed by Schmid et al. (2012) is additionally used. Schmid et al. (2012) use the partial area under the receiver operating characteristic (PAUC) for disease classification and biomarker selection as a loss function. The step-function is approximated with a sigmoid function $K(u) = 1/(1 + \exp(-u/\sigma))$, where the σ is a tuning parameter for the smoothness of the approximation. Since the AUC is a special case of the proposed function, this offers the opportunity to vary the approximation with σ. Schmid et al. (2012) propose $\sigma = min\{n_0, n_1\}^{-1/4}$ (with n_0 non-defaults and n_1 defaults) as reasonable to set σ.

7.3.2 Boosting AUC in Credit Scoring

Tree-based base learner. The evaluation for the AUC loss in boosting algorithms for credit scoring is begun with tree-based base learners. The tuning parameters maxdepth, minsplit, and the number of iterations are varied equivalent to the boosting algorithms in Section 7.2.1 to examine different combinations. Computational reasons restrict the survey to the training sample with 50% default rate. The algorithm is trained on the training sample and applied to the test and validation samples. Various combinations of tuning parameters are investigated to find the optimal values.

Table 7.10 shows an extract of the results of the analysis of the variables with coarse classification.

AUC	lowCI95%	upCI95%	iterations	maxdepth	minsplit
0.6955	0.6859	0.7052	10	1	0
0.7014	0.6917	0.7110	10	2	0
0.7088	0.6992	0.7184	10	3	0
0.7058	0.6962	0.7154	10	4	0
0.7088	0.6992	0.7184	150	3	10
0.7058	0.6962	0.7154	150	4	10

Table 7.10: Boosting results with the AUC as a loss function trained on the 50% training sample and the test data with classified variables.

The first rows are investigated with 10 iterations and an increasing maxdepth value, indicating an advancement in discriminative power for the first 3 maxdepth values. In-

terestingly, the influence of the number of iterations appears quite low for this algorithm. With a maxdepth value of 4, the outcome could not be improved with a high number of iterations. The AUC measure indicates the same value (0.7058) for 10 and 150 iterations. This phenomenon was first documented for the HingeBoost by Wang (2011).
Since the iteration number is the most important tuning parameter (compare Section 7.1.2), 10-fold cross-validated estimates of the empirical risk was used to determine the stopping iteration. Figure 7.6 shows the number of iterations on the x-axis, while the y-axis displays the (1−AUC)-loss.

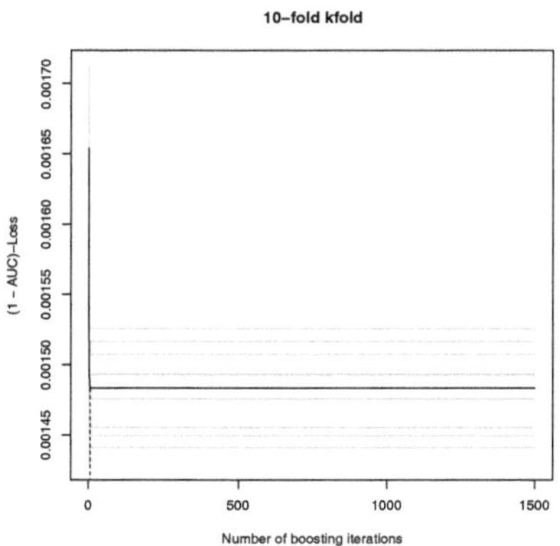

Figure 7.6: 10-fold cross validation to extract the appropriate number of boosting iterations for boosting with tree base learner and the AUC as a loss function. The optimal iteration number is 6.

The appropriate number of 6 iterations (maxdepth 4, minsplit 10) approves the previous findings that an increasing number of iterations does not go along with an increase in predictive accuracy. As seen in Figure 7.6, the empirical risk remains constant after a few iterations. The appropriate stopping iteration number determined with the 25-fold bootstrap method is 9.

Table 7.11 contains the results for the best tuning parameters for the classified variables.

		AUC	*p*-value	H measure	KS	MER
test	AUCBoost	0.7088		0.1305	0.3031	0.0263
	LogReg	0.7211		0.1421	0.3255	0.0263
validation	AUCBoost	0.7126		0.1283	0.3100	0.0260
	LogReg	0.7275	0.0618	0.1473	0.3364	0.0260

Table 7.11: Further measures for the AUCBoost trained on the 50% training sample and applied to the test and validation samples with 10 iterations, a maxdepth value of 3, and a minsplit value of 0 for classified variables. The *p*-value result from DeLong's test on the AUC measures.

All the performance measures prove the superiority of the logit model compared to the component-wise gradient boosting with AUC as a loss function. For instance, the H measure on the test data indicates a value of 0.1305 for the AUCBoost, and 0.1421 for the logit model. Note that the estimated coefficients have no probabilistic interpretation (Hothorn et al., 2012). Therefore, the Brier score is not considered in this context. The outcomes on the validation data approve the findings.

To reassess the results, the variables are tested separately to assure that the outcomes are not biased by a single variable. A high number of iterations like 500 or 1000 did not improve the results. Different step sizes of ν (for example 0.2) did not even improve the predictive performance. This coincides with the minor importance of the tuning parameter ν stated in Bühlmann and Hothorn (2007).

The analysis with the AUC as a loss function is attempted for continuous variables and results in similar conclusions. The outcomes of the AUCBoost with tree base learner are definitely below those of the logit model. Therefore, these results are not presented in detail. As described above, the PAUC function (Schmid et al., 2012) is used to vary the approximation to the jump function. A small value of σ represents a close approximation. On the test and training data, the parameter indicates only a slight influence for the results with continuous covariates. The results still achieve AUC values around 70% (compare Table A.12). In general, it follows that the idea to improve the performance of the scoring models with the AUC as a loss function in boosting methods cannot be approved for the presented credit scoring data.

Linear base learners. Within the component-wise gradient boosting, the AUC loss is also presented in combination with component-wise linear least squares as base procedures. Instead of using the negative binomial log-likelihood for boosting a linear logistic regression, the AUC loss is used for the optimization. 22 continuous and 4 categorical variables are used with the 50% training sample and data sets with encoded meaning were reduced. For the step length factor $\nu = 0.1$ is used. Table 7.12 displays the results for different iteration numbers.

test			validation			
AUC	lowCI95%	upCI95%	AUC	lowCI95%	upCI95%	iterations
0.6896	0.6796	0.6996	0.6978	0.6840	0.7116	10
0.7035	0.6936	0.7134	0.7123	0.6987	0.7260	50
0.7072	0.6973	0.7171	0.7155	0.7019	0.7291	100
0.7080	0.6981	0.7179	0.7160	0.7023	0.7296	150
0.7078	0.6979	0.7177	0.7158	0.7021	0.7294	200
0.7079	0.6980	0.7178	0.7156	0.7019	0.7292	500
0.7079	0.6980	0.7178	0.7155	0.7019	0.7292	1000

Table 7.12: Boosting linear base learners with AUC loss trained on the 50% training sample, applied to the test and validation samples.

Compared to the predictive performance of 0.7207 for the logit model with classified variables on the test data, the boosting results underachieve. The AUC values are slightly better than the logit model with continuous variables (0.7051). The same conclusion can be drawn from the validation data. The presented boosting outcomes are below the results for the logit model with 0.7224 for the validation (classified covariates). The corresponding AUC value for the logit model with continuous covariates is 0.7125.

Table 7.13 shows different performance measures for boosting the AUC with linear base learners with 148 iterations. This optimal iteration number was evaluated with cross-validation. The test results indicate different conclusions. The AUC, KS and MER show better performance for the logit model with continuous covariates compared to the AUCBoost. The H measures indicate better performance for the AUCBoost. Regarding the different measures on the validation data, the performance is better for the AUCBoost. The difference of the AUC measures, however, is not significant according to DeLong's test.

		AUC	*p*-value	H measure	KS	MER
test	AUCBoost	0.7078		0.1298	0.3063	0.0275
	LogReg	0.7051		0.1211	0.3065	0.0274
validation	AUCBoost	0.7157		0.1361	0.3179	0.0272
	LogReg	0.7125	0.6938	0.1276	0.3142	0.0272

Table 7.13: Different measures for the AUCBoost with linear base learners trained on the 50% training sample with 148 iterations and applied to the test and validation samples compared to the logit model with continuous covariates. The *p*-value results from DeLong's test on the AUC measures.

Compared to the logit model results with classified variables, the previous findings for the AUC measure are approved for the different performance measures. The H measure denotes 0.1411 on the validation data compared to 0.1361 for the AUCBoost. The KS denotes 0.3305 for the logit model with coarse classification in comparison to 0.3179 for the AUCBoost. Therefore, the AUCBoost with linear base learners is not definitely better than the classical logit model.

In Section 7.2.2, AUC values of 0.7113 for the test sample and 0.7187 for the validation sample (iteration number 379), were achieved for the boosted linear logistic regression model. This implies that the results with AUC loss do not improve the predictive accuracy compared to mentioned boosting algorithm.

The appropriate iteration number determined with cross-validation (25-fold bootstrap compare Table A.12) is 148 with AUC values of 0.7078 (test sample) and 0.7157 (validation sample). The model contains 16 covariates and outperforms neither the logit model nor the boosted logistic regression model.

Base learners with smooth effects. To add more flexibility to the model and in analogy with boosting a generalized additive model, the AUC loss is estimated with P-spline base procedures. 22 covariates are tested as smoothed effects, while 4 variables remain categorical effects. Still the 50% training sample is used to train the algorithm, followed by the application to the test and validation samples. Data with encoded meaning in the same variable were excluded (e.g. retirees). The step length factor remains 0.1. The results for different iteration numbers are shown in Table 7.14.

test			validation			
AUC	lowCI95%	upCI95%	AUC	lowCI95%	upCI95%	iterations
0.6900	0.6800	0.7000	0.6976	0.6838	0.7114	10
0.7139	0.7040	0.7237	0.7217	0.7081	0.7353	50
0.7216	0.7118	0.7314	0.7289	0.7154	0.7424	100
0.7244	0.7147	0.7342	0.7308	0.7173	0.7442	150
0.7259	0.7161	0.7356	0.7314	0.7179	0.7449	200
0.7275	0.7178	0.7373	0.7306	0.7171	0.7441	500
0.7277	0.7179	0.7374	0.7297	0.7162	0.7432	1000

Table 7.14: Boosting P-spline base learners with AUC loss trained on the 50% training sample, applied to the test and validation samples.

As shown in Table 7.14, the AUC values exceed the predictive performance of the logit model with coarse classification (0.7207 for the test sample and 0.7224 for the validation sample) with a high number of iterations. With the k-fold cross-validation that has 10 replicates, the optimal stopping iteration is determined to be 384. Compared to the tree-based base learner with AUC loss, this denotes a relatively high iteration number, and implies in contrast to the former mentioned algorithm, that the predictive accuracy increases with an increasing number of iterations. The improvement with higher iteration numbers however is slight, and the AUC value for 10 iterations is quite high as well. Moreover, the model with 384 iterations includes 25 base-learners and, therefore, a high number of covariates. The AUC values are 0.7273 (test sample) and 0.7310 (validation sample). The optimal stopping iteration with the 25-fold bootstrap method also is determined to 58 (Figure A.13). The predictive accuracy with 0.7154 and 0.7234 (test and validation sample) does not exceed the performance of the logit model. 16 base-learners are included in the model with 58 iterations.

In comparison to the boosted generalized additive model (compare Section 7.2.3), the predictive performance with AUC loss exceeds on the test sample (0.7273 to 0.7245), but remains on the same level on the validation sample (0.7310 to 0.7311) for 384 iterations. In contrast, the predictive accuracy for the model with 58 iterations underachieve the results for the boosted generalized additive model (test sample 0.7154 to 0.7245 and validation sample 0.7234 to 0.7311). As a result, a definite improvement to the boosted generalized additive model can not be concluded.

Regarding different performance measures, the same conclusions can be drawn. Table 7.15 contains the comparison of the AUCBoost with smooth effects for 58 and 384 iterations to the logit model with continuous covariates. Both boosting results with different iteration numbers outperform the logit model. The AUC difference on the validation data is significant for the AUCBoost with 384 iterations and the logit model according to DeLong's test.

		iteration	AUC	*p*-value	H measure	KS	MER
test	AUCBoost	58	0.7154		0.1372	0.3181	0.0275
	AUCBoost	384	0.7273		0.1470	0.3318	0.0275
	LogReg		0.7051		0.1211	0.3065	0.0275
validation	AUCBoost	58	0.7234	0.1824	0.1404	0.3264	0.0272
	AUCBoost	384	0.7310	0.0226	0.1497	0.3405	0.0272
	LogReg		0.7125		0.1276	0.3142	0.0272

Table 7.15: Different measures for the AUCBoost with smooth effects trained on the 50% training sample with 58 and 384 iterations and applied to the test and validation samples in comparison to the logit model with continuous covariates. The *p*-values result from DeLong's test on the AUC measures.

The comparison with different performance measures of the AUCBoost and the logit model with coarse classification indicates, however, that only a high iteration number shows better performance for the AUCBoost. The H measure of 0.1411 for the logit model with classified covariates (validation) can be outperformed with 384 iterations (0.1497) but not with 58 iterations (0.1404).

For the credit scoring case, the predictive accuracy can not be improved with the optimization concerning AUC. While the AUC loss outperforms in some cases the logit model, no advancements exist compared to the former estimated boosting algorithms presented in Section 7.2. Additionally, it seems for the analysis here that the influence of the loss function is of minor importance rather than the choice of the base-learners.

7.4 Discussion of Boosting Methods in Credit Scoring

Boosting methods are evaluated in this section for the credit scoring data. The boosting framework offers a great variety of different algorithms and shows competitive results for the scoring models compared to the classical models estimated with logistic regression.

The original AdaBoost is first applied with the Discrete, Real and Gentle AdaBoost in combination with classification trees as base learners. Most of the results outperform the classical logit model in terms of predictive accuracy measured with the AUC. A main disadvantage of this approach is the lack of interpretability. As previously discussed in the chapter on random forests, black box models are difficult to establish in the retail banking area.

The presented component-wise gradient boosting provides a huge framework of different base learners and loss functions, leading to an exhaustive amount of possible analyses. The fact that each variable is fitted separately against the gradient denotes variable selection within the algorithm, which indicates the main advantage of this method. However, the boosting methods have some weaknesses, like the lack in inference (cf. Robinzonov (2013) for further problems). The results of boosting logistic regression show good predictive performance regarding the AUC measure, even if models with many covariates arise. Promising results can be shown for boosting generalized additive models where smoothing effects are included in the models. The outcomes shown for this method, outperform the classical logit model as well as the classical additive logit model. In addition to the good performance, the component-wise gradient boosting offers the possibility to interpret the model in the final iteration. Of course, this depends on the base learner. Moreover, partial effects can be visualized for the algorithm.

Since the AUC is the most commonly used measure in credit scoring, it is used in the boosting framework as a loss function. Different base learners are combined with the AUC loss. The performance of the boosting algorithms with AUC loss and tree base learner indicates low predictive accuracy. The best predictive performance regarding the AUC measure is yielded with smoothing effects. This implies that the adequate choice of the base learner is more important for the credit scoring data than the selection of the loss function.

A critical aspect for the presented boosting algorithms is that the methods are very intensive computationally and restrict some of the investigations to the smallest training sample. Improvements in the efficiency of the algorithms are important for the implementation in the credit scoring area and the application to huge data sets. Another point for using boosted scoring models in the retail banking sector is business rule engines must be able to convert and implement the rules of the models.

As previously mentioned for recursive partitioning algorithms, boosting methods are also not implemented in many standard statistical software used in the banking sector. This makes it difficult to analyze and test methods easily in real-life practice. Nevertheless, the presented credit scoring results evaluated with different boosting algorithms give promising results for further research and effort to implement and apply boosting algorithms in scoring models.

Chapter 8

Summary and Outlook

Improving credit scoring models in terms of predictive accuracy is the main focus of this thesis. Recent methods from statistics and machine learning are benchmarked regarding the AUC measure. Figure 8.1 gives an overview of the algorithms evaluated for the credit scoring data and Table 8.1 shows a comparison of the methods for the 50% training sample and continuous covariates. The thesis covers classical models like logistic regression and generalized additive models, an AUC approach, recursive partitioning methods, and finally, boosting algorithms from machine learning.

As a starting point, the AUC measure is explained in Chapter 2 and further performance measures, used in credit scoring, are presented. A description of the German retail credit data is given as a basis for the presented evaluation.

The classical scorecard development process is briefly presented in Chapter 3 to give some insights in the scoring models estimated and used in practice. Besides the obligatory data preparation, the standard procedure covers the univariate and multivariate analyses, where the variables are investigated separately regarding their predictive performance and in combination in the logistic regression model. The estimates resulting from the final model are normally converted to so-called scores. This is just a scaling to a specific range 1 to 1000, for example, which does not change the ranking order of the observations, but provides easily interpretable models that are suitable for straightforward implementation in the business rule engine.

Chapter 4 is concerned with the AUC approach, where the widely applied AUC measure is introduced as direct optimization criterion. The relationship of the AUC to the Wilcoxon statistic is outlined and the Nelder–Mead algorithm is used as derivative-free optimization procedure, because the AUC is as a sum of step functions not differentiable. The properties of this approach are analyzed within a simulation study where five different link functions are simulated in the data to compare the AUC approach and the logit model. The results indicate the tendency that the AUC approach yields better predictive accuracy if the link function is not logistic. This implies that the AUC approach outperforms in cases where the distribution assumption of the logit model fails. In these cases, the optimality properties of the AUC approach still hold for monotone increasing functions. In the AUC approach, the coefficients of the classical logit model are used as starting values for the optimization.

The evaluation of the credit scoring data shows slight improvements for the AUC approach compared to the logit model.

The credit scoring problem in this thesis is also evaluated with a Generalized Additive Model (Chapter 5). By including nonlinearities in the algorithm, the logistic additive model indicates higher predictive accuracy compared to the classical logit model. This approach offers interesting insights in the relationship of the dependent variable and the covariates and should at least be considered as supplementary method for analyzing credit defaults.

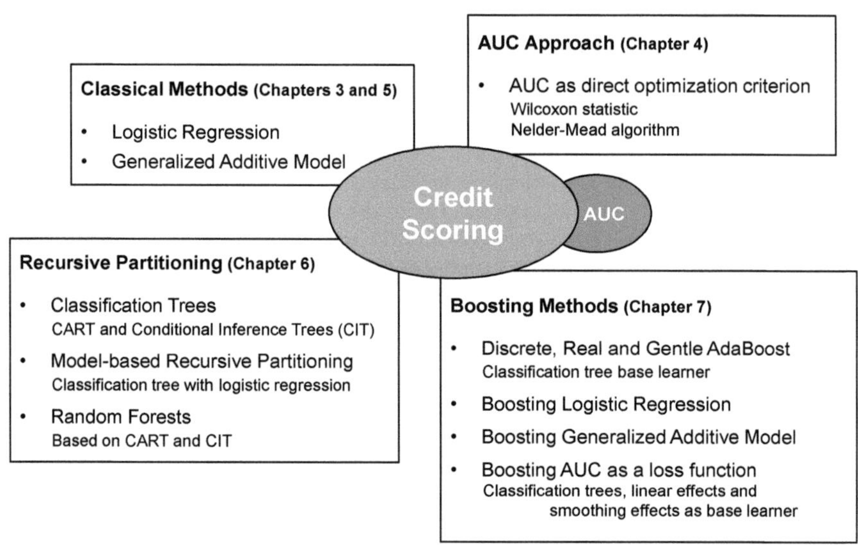

Figure 8.1: Overview of the presented methods in this thesis for the credit scoring evaluation.

A main part of this thesis is concerned with the evaluation of recursive partitioning algorithms in credit scoring (Chapter 6). The presented conditional inference trees overcome the variable selection bias of the original CART algorithm, but both approaches lead to very poor predictive performance for the credit scoring data even if the binary structure and the clarity of the classification rules would imply an advantage.

An interesting approach that shows competitive results compared to the logit model for the credit scoring data, is the model-based recursive partitioning. Testing the stability of the coefficients is an interesting approach and can help improving scoring models. The complexity of the models should be taken into account.

The performance concerning the AUC measure is different for both proposed random forest algorithms. Random forests overcome the instability of single classification trees and outperform them in terms of the AUC measure, but they are difficult to interpret due to their structure. The results for the random forests with conditional inference trees are better than those based on CART like trees. Even if random forests indicate high predictive accuracy for the validation data and the 50% training sample shown in Table 8.1, random forest algorithms in the presented analysis outperform in some but not in all cases shown in Chapter 6. In other scientific fields, and even in some other credit scoring studies, good performances are confirmed for random forest algorithms. The variable importance measures are useful tools for supporting the classical scorecard development process.

Different methods	AUC measure	
Discrete AdaBoost	0.7275	Chapter 7/ Table A.8
Boosting GAM	0.7267	Chapter 7/ Table 7.9
GAM	0.7263	Chapter 5/ Table 5.1
Gentle AdaBoost	0.7249	Chapter 7/ Table 7.2
Random forest CIT	0.7241	Chapter 6/ Table 6.18
Real AdaBoost	0.7236	Chapter 7/ Table A.6
AUCBoost smooth effects	0.7234	Chapter 7/ Table 7.15
Boosting LogReg	0.7182	Chapter 7/ Table 7.6
Random forest CART	0.7163	Chapter 6/ Table 6.14
AUCBoost linear base learner	0.7157	Chapter 7/ Table 7.13
AUC Approach	0.7141	Chapter 4/ Table 4.5
Model based tree	0.7134	Chapter 6/ Table 6.11
Logistic regression	0.7125	Chapter 3/ Table 5.1
AUCBoost tree base learner	0.6993	Chapter 7/ Table A.12
CART	0.6814	Chapter 6/ Table 6.3
Conditional inference tree	0.6756	Chapter 6/ Table 6.6

Table 8.1: Comparison of the presented methods in this thesis for the validation data, trained on the 50% training sample with continuous covariates. The corresponding AUC measure for the logit model with classified variables denotes 0.7224.

Another principal part of this thesis is Chapter 7 that includes the evaluation of boosting methods from machine learning for credit scoring. Good results and competitive performance can be stated for the predictive accuracy of some boosting algorithms compared to the classical scoring model. The Gentle AdaBoost, with classification trees as base learner, represents outperforming outcomes, compared to the logit model with respect to the AUC measure. This approach represents a kind of black box model, and is, therefore, critical for the application in credit scoring. For the comparison in Table 8.1, the best AUC measure is shown for Discrete AdaBoost. The boosting of generalized additive models shows improvements in performance compared to the classical logit model, as well as compared to the additive logistic regression. Since this approach belongs to the component-wise gradient boosting, variable selection is included in the algorithm and more views into the results are possible.

While other presented boosting methods, like boosting logistic regression, provide competitive results with respect to the logit model, the boosting algorithms with the AUC as a loss function yield AUC values below the corresponding classical scoring models with tree base learners. By using the AUC as a loss function in combination with smooth effects or linear base learners, the predictive performance is quite competitive compared to the logit model.

This indicates that the choice of the base learner is more important in this case than using the AUC as a loss function.

To summarize, the main conclusions of my thesis are as follows. The classical scoring model, based on logistic regression, still provides robust and interpretable results with good predictive performance. Using the AUC as direct objective function is a competitive alternative, since the AUC measure still dominates in credit scoring in terms of measuring predictive accuracy. Specific boosting algorithms improve the performance of classical scoring models regarding the AUC measure. Recursive partitioning methods can support the development of scoring models with interesting analytical tools. As shown in the comparison of the methods for the 50% training sample in Table 8.1, the newer presented methods indicate competitive predicitive performance compared to the classical logit model.

Relevant issues in retail credit scoring are outlined with some remarks for further research:

Interpretability. As mentioned in this thesis, the interpretability of the scoring models and the ease of explanation are main features in the credit scoring context. The acceptance of methods in the retail banking practice depends on the complexity of the algorithms. This is, in fact, a tradeoff between the desire for the highest achievable peformance and the complexity of the models. The presented boosting methods offer better predictive performance than the classical scoring models but represent more complex algorithms. There is, however, an ongoing process for developing and enhancing these algorithms that could lead to more application in the banking sector.

Feasibility and Implementation. For the application of new statistical methods in credit scoring practice, the first step is the availability of the methods in the standard statistical software packages used in practice. Vice versa, the use of the R Software in the banking area would be a possibility to open the methodological view of scoring models on a broad base. Another task deals with the implementation of the new methods in business rule engines. An important point not covered in this thesis is converting the methods into rules feasible for implementation.

Developments in Banking. This point deals with a commercial point of view, rather than a statistical one. New regulations in the banking sector, for example, resulting from Basel II (Basel Committee on Banking Supervision, 2004) have a permanent influence on scoring models. The definition of default is a case of how requirements can affect the development process. This is especially relevant for behavioral scoring, which is not evaluated in this thesis since the focus is on the application scoring. Whether the recent algorithms presented in this work could improve the performance of the behavioural scoring models, is an interesting question for further analyses.
Another point in this context are data protection regulations, which can influence the utilization of the variables apart from statistical performance analyses. Additionally, legislation can affect the use of specific covariates. This is already the case in the insurance

business, where gender can not be used anymore for the risk assessment, and unisex rates are required.

Statistical Developments. Some of the newer presented methods are very intensive computationally and restrict the analyses in some parts to the smallest training sample. New developments regarding efficiency of the algorithms could lead to more popularity in credit scoring. The application to big data is an important aspect even in many other scientific fields. Increasing computational processors combined with developed software offer new possibilities to implement new classification rules in the credit scoring environment. Moreover, the newer methods still indicate problems like a lack in inference for boosting algorithms. Further research for more insights and interpretability would also be promising for the application in the retail banking sector.
High accuracy in scoring models will always be a key issue for retail credit banks. Consequently, the method comparison of statistical state-of-the-art algorithms continues to be an essential question in credit risk assessment. Breiman (2001b, p.199) states that for solving problems with real world data "...We need to move away from exclusive dependence on data models and adopt a more diverse set of tools".

Appendix A

Supplementary Material

A.1 Optimization AUC - Chapter 4

The following Figures (A.1 to A.4) result from the simulation study in Chapter 4 and denote boxplots to compare the AUC approach and the logit model. 100 simulations were run for the simulation and different link functions were simulated in the data. The procedure is described in Section 4.2.2.

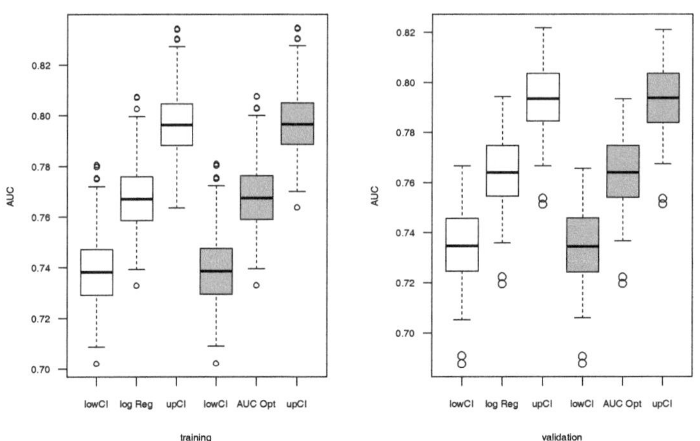

Figure A.1: AUC comparison for the logit model (white) and the AUC approach (grey) with confidence intervals on the simulated training data and applied to validation data with logit link.

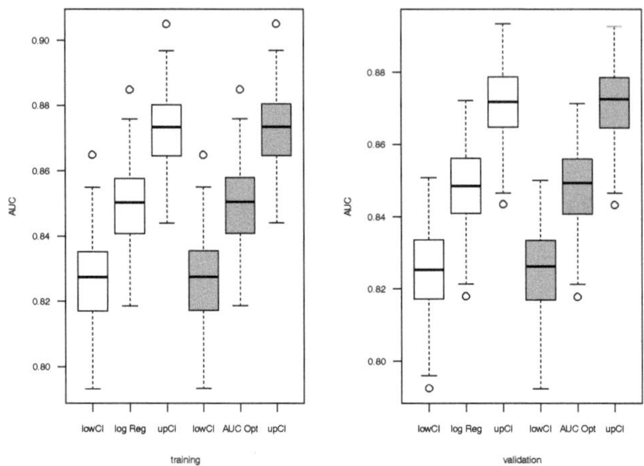

Figure A.2: AUC comparison for the logit model (white) and the AUC approach (grey) with confidence intervals on the simulated training data and applied to validation data with complementary log log link.

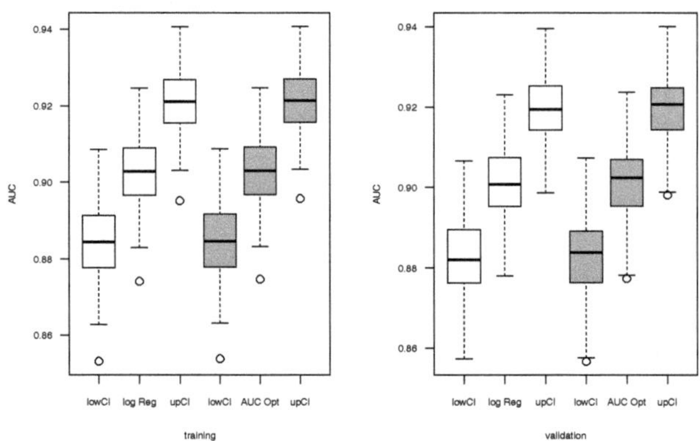

Figure A.3: AUC comparison for the logit model (white) and the AUC approach (grey) with confidence intervals on the simulated training data and applied to validation data with link1.

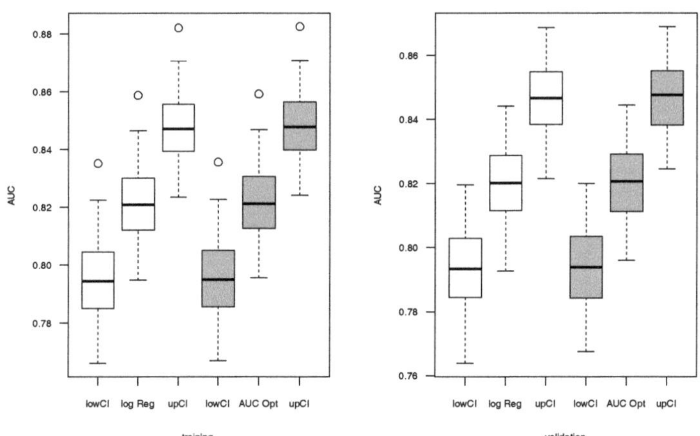

Figure A.4: AUC comparison for the logit model (white) and the AUC approach (grey) with confidence intervals on the simulated training data and applied to validation data with link2.

A.2 Generalized Additive Model - Chapter 5

The following table shows the comparison of the generalized additive models to the logit model with classified covariates.

		AUC	Gini	H measure	KS	MER	Brier score
trainORG	GAM	0.7250	0.4501	0.1404	0.3338	0.0272	0.0270
	LogReg	0.7219	0.4437	0.1415	0.3247	0.0272	0.0271
train020	GAM	0.7250	0.4500	0.1403	0.3313	0.0271	0.0703
	LogReg	0.7237	0.4474	0.1428	0.3285	0.0272	0.0703
train050	GAM	0.7263	0.4527	0.1414	0.3314	0.0272	0.2353
	LogReg	0.7224	0.4448	0.1411	0.3305	0.0272	0.2351

Table A.1: Further measures for the GAM results compared to the logistic regression results with classified variables for the validation data.

A.3 Recursive Partitioning - Chapter 6

The following tables show results of classification trees based on the CART algorithm referenced in Section 6.4.1. Table A.2 includes the information split criterion, Table A.3 shows outcomes for different prior probabilities and the Tables A.4 and A.5 contain results of trees with m-estimate smoothing.

train050	AUC	minsplit	cp-value	number of splits	split criterion	AUC LogReg
test	0.6707	10	0.00395778	16	info	0.7051
validation	0.6771	10	0.00395778	16	info	0.7125

Table A.2: Prediction accuracy for the CART classification trees trained on the 50%-training sample, applied to the test and validation samples for the information index as split criterion.

test			validation			
AUC	lowCI95%	upCI95%	AUC	lowCI95%	upCI95%	priors
0.6679	0.6578	0.6780	0.6627	0.6487	0.6767	without
0.6340	0.6238	0.6442	0.6320	0.6179	0.6461	0.95/0.05
0.5560	0.5459	0.5661	0.5381	0.5242	0.5520	0.9/0.1

Table A.3: Prediction accuracy for CART classification trees trained on the original training sample, applied to the test and validation samples with the gini split criterion and different prior probabilities. The minsplit value is fixed to a value of 20, while the cp-value is fixed to 0.0001.

	AUC	minsplit	m	p	cp-value	number of splits	AUC LogReg
trainORG	0.6800	10	5	0.974	0.00024540	332	0.7045
train020	0.6772	20	20	0.974	0.00085890	144	0.7054
train050	0.6734	100	5	0.974	0.00043975	28	0.7051

Table A.4: Prediction accuracy for the best CART classification trees with m-estimate smoothing and gini index as split criterion, trained on the different training samples and applied to the test sample.

	AUC	minsplit	m	p	cp-value	number of splits	AUC LogReg
trainORG	0.6731	10	5	0.974	0.00024540	332	0.7118
train020	0.6793	20	20	0.974	0.00085890	144	0.7129
train050	0.6827	100	5	0.974	0.00043975	28	0.7125

Table A.5: Validation results for the best CART classification trees with m-estimate smoothing and gini index as split criterion, trained on the different training samples, applied to the validation sample.

Figure A.5 is referenced in Section 6.4.3 where random forests are analysed for the credit scoring case. The variable importance measure 'mean decrease in node impurity' is shown for all three training samples.

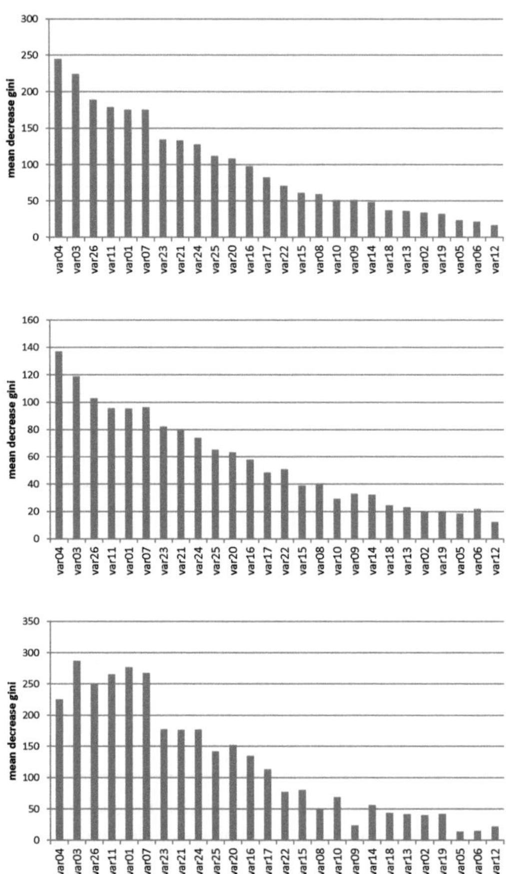

Figure A.5: Mean decrease in node impurity for the 26 covariates on the 20% training sample (top), for the 50% training sample (middle) and the original training sample (bottom).

Figure A.6 shows AUC measures for random forest results based on conditional inference trees. The number of trees is fixed to 200, while the mtry value is changed. The figure presents results with continuous covariates.

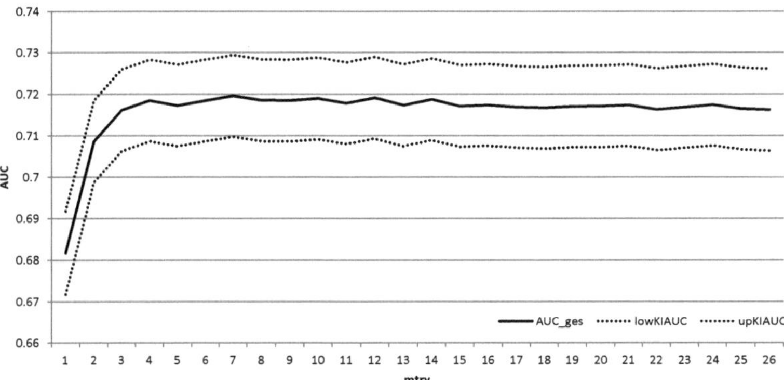

Figure A.6: Prediction accuracy for random forests based on conditional inference trees trained on the 50% training sample and applied to the test sample by varying the mtry-values for 200 trees.

A.4 Boosting - Chapter 7

The following Tables (A.6 to A.9) present results referenced in Section 7.2.1 where boosting methods are evaluated for the credit scoring data. Classification trees are used as base learners for the Discrete AdaBoost and the RealAdaBoost and continuous covariates are employed.

		AUC	iterations	maxdepth	minsplit	AUC LogReg	p-value
trainORG	test	0.7156	150	8	0	0.7045	
	validation	0.7198	150	8	0	0.7118	0.3340
train020	test	0.7170	150	30	0	0.7054	
	validation	0.7206	150	30	0	0.7129	0.3430
train050	test	0.7177	150	15	10	0.7051	
	validation	0.7236	150	15	10	0.7125	0.1726

Table A.6: Real AdaBoost results with continuous variables trained on three different training samples and applied to the test and validation samples in comparison to the logit model. The p-values result from DeLong's test.

		AUC	Gini	H measure	KS	MER	Brier score
trainORG	Real	0.7198	0.4396	0.1363	0.3239	0.0272	0.0786
	LogReg	0.7118	0.4237	0.1283	0.3101	0.0272	0.0270
train020	Real	0.7206	0.4412	0.1349	0.3241	0.0272	0.1116
	LogReg	0.7129	0.4257	0.1286	0.3143	0.0272	0.0719
train050	Real	0.7236	0.4473	0.1418	0.3259	0.0271	0.2282
	LogReg	0.7125	0.4249	0.1276	0.3142	0.0272	0.2494

Table A.7: Further measures for the Real AdaBoost results with continuous variables compared to the logistic regression results for the validation data.

		AUC	iterations	maxdepth	minsplit	AUC LogReg	p-value
trainORG	test	0.7230	150	4	0	0.7045	
	validation	0.7277	150	4	0	0.7118	0.0526
train020	test	0.7226	150	5	0	0.7054	
	validation	0.7269	150	5	0	0.7129	0.0861
train050	test	0.7244	150	5	0	0.7051	
	validation	0.7275	150	5	0	0.7125	0.0664

Table A.8: Discrete AdaBoost results with continuous variables trained on three different training samples and applied to the test and validation samples in comparison to the logit model. The p-values result from DeLong's test.

		AUC	Gini	H measure	KS	MER	Brier score
trainORG	Discrete	0.7277	0.4554	0.1450	0.3316	0.0272	0.0270
	LogReg	0.7118	0.4237	0.1283	0.3101	0.0272	0.0270
train020	Discrete	0.7269	0.4538	0.1465	0.3279	0.0271	0.0648
	LogReg	0.7129	0.4257	0.1286	0.3143	0.0272	0.0719
train050	Discrete	0.7275	0.4550	0.1487	0.3383	0.0272	0.2276
	LogReg	0.7125	0.4249	0.1276	0.3142	0.0272	0.2494

Table A.9: Further measures for the Discrete AdaBoost results with continuous variables compared to the logistic regression results for the validation data.

The following four Figures (A.7 to A.10) and Table A.10 result from evaluating the credit scoring data with boosting the logistic regression model in Section 7.2.2.
Figure A.7 contains cross-validation results and Figure A.8 shows the AUC values for the iterations from 1 to 1000. Table A.10 and the Figures A.9 and A.10 provide the test and validation results for boosting logistic regression with categorical effects and iteration numbers from 1 to 1500.

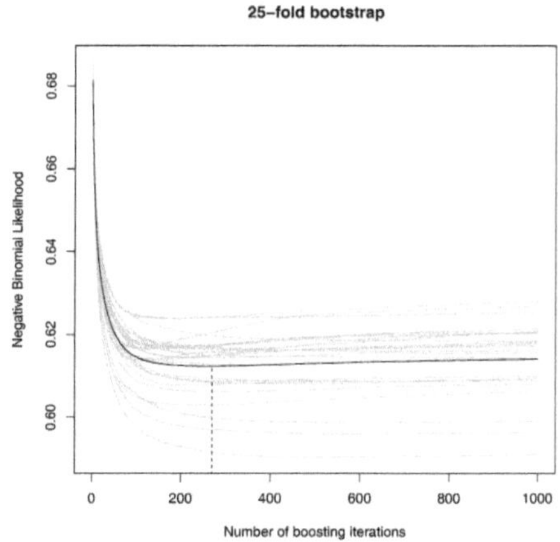

Figure A.7: Boosting logistic regression - cross validation with 25-fold bootstrapping to extract an appropriate iteration number (270).

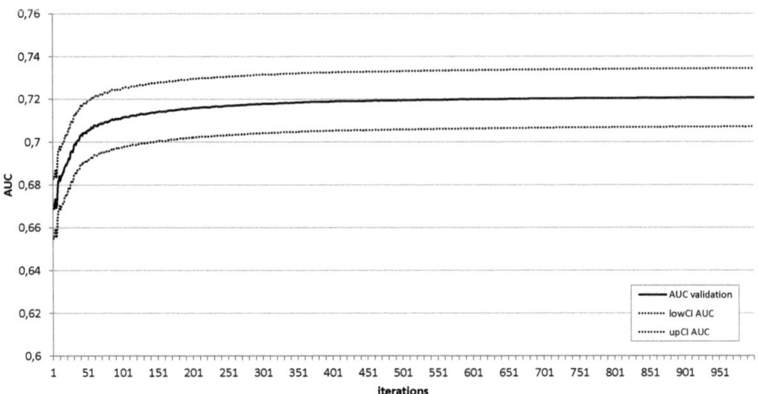

Figure A.8: AUC values with confidence intervals for boosting logistic regression on the validation sample for iterations 1 to 1000 with 26 potential continuous variables.

test			validation			
AUC	lowCI95%	upCI95%	AUC	lowCI95%	upCI95%	iterations
0.6510	0.6412	0.6608	0.6632	0.6496	0.6768	10
0.6940	0.6843	0.7036	0.7047	0.6913	0.7180	50
0.7069	0.6973	0.7165	0.7179	0.7047	0.7312	100
0.7118	0.7022	0.7214	0.7218	0.7086	0.7351	150
0.7236	0.7141	0.7331	0.7313	0.7182	0.7444	500
0.7266	0.7171	0.7360	0.7334	0.7203	0.7165	1000
0.7269	0.7175	0.7364	0.7335	0.7204	0.7466	1200
0.7273	0.7178	0.7368	0.7335	0.7204	0.7466	1500

Table A.10: Boosting logistic regression trained on the 50% training sample, applied to the test and validation samples with categorical variables.

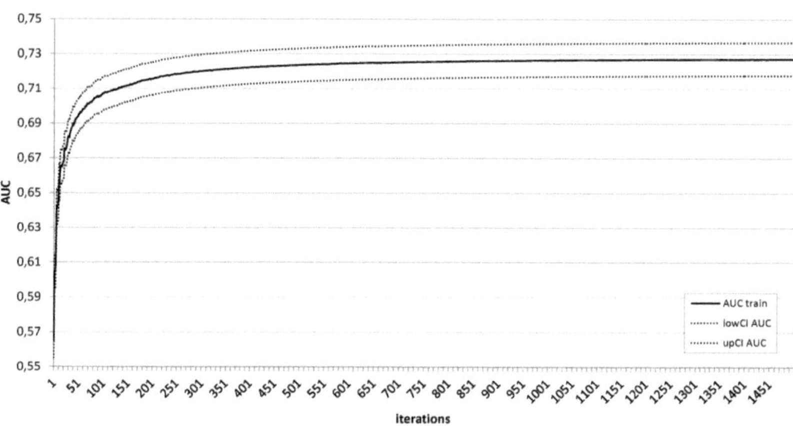

Figure A.9: AUC values with confidence intervals for boosting logistic regression on the test sample for iterations 1 to 1500 with 26 potential categorical variables.

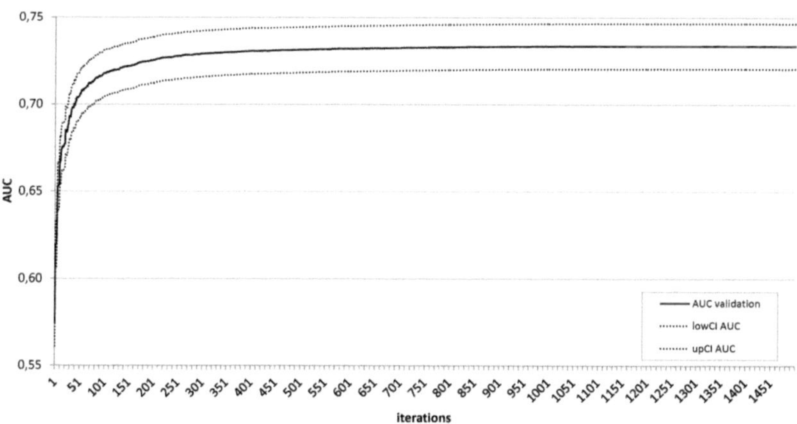

Figure A.10: AUC values with confidence intervals for boosting logistic regression on the validation sample for iterations 1 to 1000 with 26 potential categorical variables.

Figure A.11 and Table A.11 are referenced in Section 7.2.3. The outcomes result from boosting a generalized additive model for the credit scoring data. While Figure A.11 contains cross validation results, Table A.11 shows the prediction accuracy for boosting GAMs with 8 potential variables.

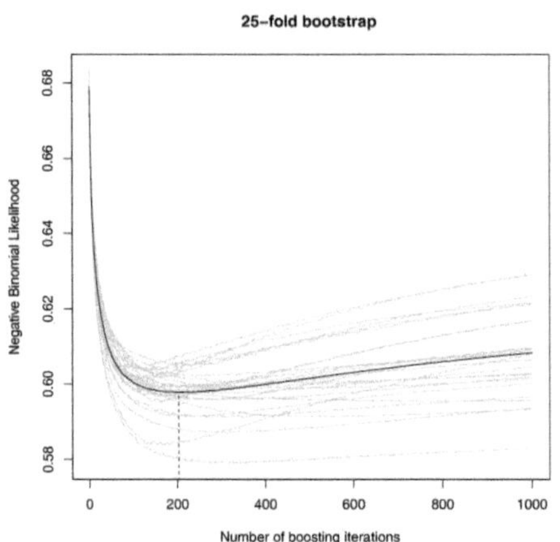

Figure A.11: Boosting generalized additive models - cross validation with 25-fold bootstrapping to extract an appropriate iteration number (203).

test			validation			
AUC	lowCI95%	upCI95%	AUC	lowCI95%	upCI95%	iterations
0.6769	0.6668	0.6869	0.6845	0.6706	0.6984	10
0.7098	0.6999	0.7197	0.7171	0.7035	0.7308	50
0.7175	0.7077	0.7273	0.7245	0.7109	0.7380	100
0.7200	0.7102	0.7298	0.7261	0.7126	0.7397	150
0.7213	0.7115	0.7311	0.7267	0.7131	0.7402	211
0.7215	0.7117	0.7313	0.7266	0.7131	0.7402	233
0.7219	0.7121	0.7317	0.7266	0.7131	0.7402	261
0.7226	0.7128	0.7324	0.7258	0.7123	0.7393	500
0.7223	0.7125	0.7321	0.7242	0.7107	0.7378	1000

Table A.11: Boosting generalized additive models with P-spline base-learners and the negative binomial log-likelihood loss trained on the 50% training sample, applied to the test and validation samples with 8 potential variables.

In Section 7.3.2, the credit scoring data are evaluated with boosting algorithms and the AUC as a loss function. Figure A.12 and Figure A.13 are referenced there, and show results for the cross-validation to find the optimal iteration number.

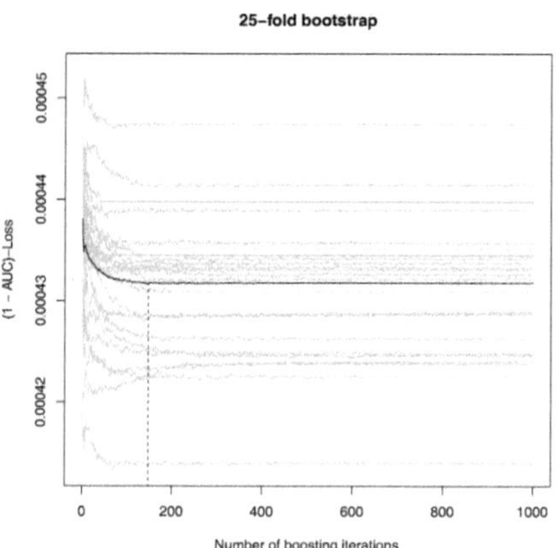

Figure A.12: Boosting linear effects with AUC loss function - cross validation with 25-fold bootstrapping to extract an appropriate iteration number (148).

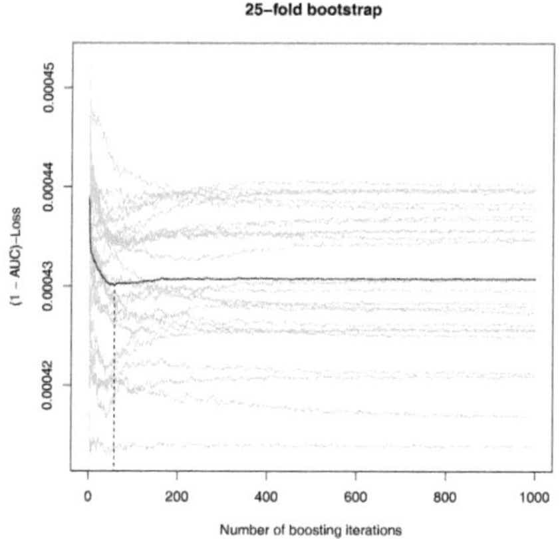

Figure A.13: Boosting smooth effects with AUC loss function - cross validation with 25-fold bootstrapping to extract an appropriate iteration number (58).

Table A.12 is referenced in section 7.3.2 and contains AUC measures for the boosting method applied to the credit scoring data with the AUC as a loss function.

test			**validation**						
AUC	lowCI95%	upCI95%	AUC	lowCI95%	upCI95%	iterations	maxdepth	minsplit	sigma
0.6854	0.6757	0.6951	0.6870	0.6735	0.7005	100	4	0	0.05
0.6949	0.6852	0.7046	0.6993	0.6859	0.7127	100	4	0	0.1
0.6993	0.6897	0.7090	0.7033	0.6900	0.7167	100	4	0	**0.16**
0.7045	0.6949	0.7141	0.7083	0.6950	0.7216	100	4	0	0.2
0.7096	0.7001	0.7192	0.7092	0.6959	0.7226	100	4	0	0.3
0.7054	0.6958	0.7150	0.7041	0.6908	0.7175	100	4	0	0.5
0.6866	0.6769	0.6963	0.6839	0.6703	0.6974	100	4	0	0.8

Table A.12: Boosting results with AUC as a loss function for the test and validation sample trained on the 50% training sample with different values for sigma as approximation to the jump function. The smaller the σ the closer the approximation to the jump function.

Appendix B

Computational Aspects

B.1 Statistical Software

The R system (R Development Core Team, 2012) for statistical computation and the statistical analysis software SAS were the basis for all analyses presented in this thesis. In the following, add-on R packages are highlighted, which were mainly used in the investigations. For standard procedures like GLMs or the creation of figures, I refer to the standard manuals of the R language.

The R package `optimx` (Nash and Varadhan, 2012) contains an implementation of the Nelder–Mead method, which was used in Chapter 4 for the AUC optimization. An example of the R-Code for the AUC approach and the corresponding simulation study is also presented in the following Section B.2.

The analyses for the Generalized Additive Models in Chapter 5 were performed with the R package `gam` (Hastie, 2011). The GAM framework of Wood (2006), which is referenced for an outlook, can be investigated with the R package `mgcv` (Wood, 2013).

The classification trees based on CART in Chapter 6 were performed with the R package `rpart` (Therneau and Atkinson, 2010). The R package `party` (Hothorn et al., 2013) was used for the investigations of conditional inference trees. The traditional random forests are implemented in the R package `randomForest` of Liaw and Wiener (2009), while the random forests based on conditional inference trees are included in the R package `party` (Hothorn et al., 2013). The results for the model-based recursive partitioning algorithms in Chapter 6 were also produced with the R package `party`.

In Chapter 7, the R packages `mboost` (Hothorn et al., 2012) and `ada` (Culp et al., 2007) were used as implementations for the different boosting techniques. Moreover, the R-Code provided by Schmid et al. (2012) for the PAUC function was used in the boosting chapter.

B.2 Selected R-Code

In addition to the statistical software details in the former section (Appendix B.1), examples for the AUC approach and the simulation study of Chapter 4 are presented here.

The following R-Code gives an example of the AUC approach for three explanatory variables and a binary outcome variable to illustrate the procedure.

```
# AUC approach - example 3 variables
# optimization beta coeff with AUC und Nelder Mead
# AK, 30.03.2012

#load the required packages
    library(optimx)
    library(reshape)

##step 1: function to create data for AUC with Wilcoxon
# data: data sample
# id: unique id
# gs: binary outcome variable
# var1 to var3: explanatory variables

dataufbereiten3var <- function(data, id, gs, var1, var2, var3)    {

  data1 <- data[,c(id,gs,var1,var2,var3)]  #take these vars
  Gute <- subset(data1, data1[,c(gs)]==0) #take non-defaults
  Schlechte <- subset(data1, data1[,c(gs)]==1) #take defaults

  #rename variables for non-defaults
  Gute <- rename.vars(Gute,c(id,gs,var1,var2,var3),c(paste(id,"G",sep="_"),
   paste(gs,"G",sep="_"),paste(var1,"G",sep="_"),paste(var2,"G",sep="_"),
   paste(var3,"G",sep="_")))

  #rename variables for defaults
  Schlechte <- rename.vars(Schlechte, c(id,gs,var1,var2,var3),
   c(paste(id,"S",sep="_"),paste(gs,"S",sep="_"),paste(var1,"S",sep="_"),
   paste(var2,"S",sep="_"),paste(var3,"S",sep="_")))

  #build all combinations with the id's and rename id's
  neu <- expand.grid(Gute[,c(paste(id,"G",sep="_"))],
   Schlechte[,c(paste(id,"S",sep="_"))])
  neu <- rename.vars(neu, c("Var1","Var2"), c(paste(id,"G",sep="_"),
   paste(id,"S",sep="_")))

  #add variables for non-defaults and defaults
  newdata <- merge(neu, Gute, by=paste(id,"G",sep="_"))
```

```
  newdata2_3var <- merge(newdata, Schlechte, by=paste(id,"S",sep="_"))

  #return(newdata2_3var)
  save("newdata2_3var", file="...")
    }

##step 2: function to calculate AUC with Wilcoxon
# beta: start values --> fix one value
# data: data created with former function
# var1 to var3: explanatory variables

wilkAUC_bs50 <- function (beta, data, var1,var2,var3) {
   #build the difference
   diffq <-  (1)*data[,paste(var1,"S",sep="_")]-(1)*
    data[,paste(var1,"G",sep="_")]+beta[1]*data[,paste(var2,"S",sep="_")]-
    beta[1]*data[,paste(var2,"G",sep="_")]+beta[2]*
    data[,paste(var3,"S",sep="_")]-beta[2]*data[,paste(var3,"G",sep="_")]
   #Wilcoxon
   neux <- ifelse(diffq < 0,  c(0), (ifelse(diffq == 0, c(0.5),  c(1))))
   anz2 <- nrow(data)
   summe2 <- sum(neux)
   AUC2 <- summe2/anz2  #AUC value
   return(-AUC2)
     }

##step 3: optimization betas with AUC and Nelder Mead
# par: start values from logistic regression - fix one value (normalization)
# fn: wilkAUC_bs50 from step 2
# data: data sample from step 1
# var1 to var3: explanatory variables
# alternative to optimx --> optim

optimx(c( betaLR2/betaLR1,betaLR3/betaLR1), wilkAUC_bs50,data=newdata2_3var,
method=c("Nelder-Mead"),
var1="name1", var2="name2",var3="name3")
```

The R-Code for the simulation study used in Chapter 4 is presented in the following with the example for simulating the logistic relationship in the data. The former functions for the AUC approach are used within the following R-Code.

```
#load required packages
library(reshape)
library(optimx)
library(mvtnorm)
```

```
list_fin_mult<-list()
  for (k in 1:100){

#step1a) Simulate data for the model estimation with logit function
  n=1000
  coefs=c(1,0.5,0.3)
  sigma=diag(length(coefs))
  dat<-rmvnorm(n, sigma=sigma)
  x.beta=dat %*% coefs
  dat<-as.data.frame(dat)
  names(dat)<-c("x1","x2","x3")
  dat$y<-as.numeric(rbinom(n, 1, exp(x.beta)/(1+exp(x.beta))))
  head(dat)
  schlecht_anz<-sum(dat$y)

#step1b) Simulate data for validation with logit function
  nv=1000
  coefsv=c(1,0.5,0.3)
  sigmav=diag(length(coefsv))
  datv<-rmvnorm(nv, sigma=sigmav)
  x.betav=datv %*% coefsv
  datv<-as.data.frame(datv)
  names(datv)<-c("x1","x2","x3")
  datv$y<-as.numeric(rbinom(nv, 1, exp(x.betav)/(1+exp(x.betav))))
  head(datv)
  schlecht_anz_v<-sum(datv$y)

#step2) Estimate coefficients with glm
  modlog<-glm(y ~ x1+x2+x3,data=dat, family=binomial(link="logit"))
  summary(modlog)

    #AUC measure with confidence intervals for training data
    predScore<-predict(modlog, newdata=dat, type="response")
    predl<-cbind(dat[,c("y")],predScore)
    predl<-as.data.frame(predl)
    names(predl)[1]<-"gs_m3ki_18"
    names(predl)[2]<-"Score"
    Gini_ges<-Ginifunc(predl,"gs_m3ki_18","Score")
    AUC_ges<- 1/2*Gini_ges+0.5
    q1<-AUC_ges/(2-AUC_ges)
    q2<-(2*AUC_ges^2)/(1+AUC_ges)
    anz<-nrow(dat)
    schlecht<-sum(dat[,c("y")])
    gut<-anz-schlecht
    se.auc<-sqrt((AUC_ges*(1-AUC_ges)+(schlecht-1)*(q1-AUC_ges^2)+(gut-1)*
```

```
     (q2-AUC_ges^2))/(gut*schlecht))
    lowCIAUC<-max(0,AUC_ges-qnorm(0.975)*se.auc)
    upCIAUC<-min(1,AUC_ges+qnorm(0.975)*se.auc)

#step3) Apply model on validation data and calculate AUC measure
#with confidence intervals
    predScoreV<-predict(modlog, newdata=datv, type="response")
    predlV<-cbind(datv[,c("y")],predScoreV)
    predlV<-as.data.frame(predlV)
    names(predlV)[1]<-"gs_m3ki_18"
    names(predlV)[2]<-"Score"
    Gini_ges_v<-Ginifunc(predlV,"gs_m3ki_18","Score")
    AUC_ges_v<- 1/2*Gini_ges_v+0.5
    ...
    lowCIAUC_v<-max(0,AUC_ges_v-qnorm(0.975)*se.auc)
    upCIAUC_v<-min(1,AUC_ges_v+qnorm(0.975)*se.auc)

   #coefficients of the logit model
   coefx1<-coef(modlog)["x1"]
   coefx2<-coef(modlog)["x2"]
   coefx3<-coef(modlog)["x3"]
   #result list 1
   list<-cbind(Gini_ges,AUC_ges,lowCIAUC,upCIAUC,schlecht_anz,coefx1,coefx2,
    coefx3,Gini_ges_v,AUC_ges_v,lowCIAUC_v,upCIAUC_v,schlecht_anz_v)


#step4) AUC optimization with coefficients of the logit model as starting values
    counter <- row.names(dat) #unique ID
    dat<-cbind(dat,counter)
    #use former function for data preparation
    dataufbereiten3var(dat, "counter", "y", "x1", "x2","x3")
    load("...")

    #AUC optimization with Nelder-Mead
    testb<-optimx(c( coef(modlog)["x2"]/coef(modlog)["x1"],
     coef(modlog)["x3"]/coef(modlog)["x1"]), wilkAUC_bs50Sim,
     data=newdata2_3var, method=c("Nelder-Mead"),var1="x1", var2="x2",var3="x3")
    coeffsauc<-unlist(testb[1])
    coeffx2auc<-coeffsauc[1]
    coeffx3auc<-coeffsauc[2]
    AUCopt<-testb[2]

    AUCopt<-as.numeric(AUCopt$fvalues)
    AUCopt<-AUCopt*(-1)
    ...
```

```
    lowCIAUCopt<-max(0,AUCopt-qnorm(0.975)*se.auc)
    upCIAUCopt<-min(1,AUCopt+qnorm(0.975)*se.auc)
    #result list 2
    list2<-cbind(test_AUC,coeffx2auc,coeffx3auc,AUCopt,lowCIAUCopt,upCIAUCopt)

#step5) Apply AUC optimized values for validation data and calculate AUC
#for logit model and AUC approach
      ScoreTest<- 1/(1+exp(-(coef(modlog)["(Intercept)"])-
       ((coef(modlog)["x1"])*datv$x1)-((coef(modlog)["x2"])*datv$x2)-
       ((coef(modlog)["x3"])*datv$x3)))
      ScoreAUC<- 1/(1+exp(-(1*datv$x1)-(coeffx2auc*datv$x2)-
       (coeffx3auc*datv$x3)))
      datv<-cbind(datv,ScoreTest,ScoreAUC)
      datv<- rename.vars(datv,c("y"), c("gs_m3ki_18"))
      Gini_gestest_v<- Ginifunc(datv,"gs_m3ki_18","ScoreTest")
      AUC_gestest_v<- (1/2*Gini_gestest_v)+0.5

      Gini_gesAUCopt_v<- Ginifunc(datv,"gs_m3ki_18","ScoreAUC")
      AUC_gesAUCopt_v<- (1/2*Gini_gesAUCopt_v)+0.5
      ...
      lowCIAUCopt_v<-max(0,AUC_gesAUCopt_v-qnorm(0.975)*se.auc)
      upCIAUCopt_v<-min(1,AUC_gesAUCopt_v+qnorm(0.975)*se.auc)
      #result list 3
      list3<-cbind(Gini_gestest_v,AUC_gestest_v,Gini_gesAUCopt_v,
       AUC_gesAUCopt_v,lowCIAUCopt_v,upCIAUCopt_v)

      #combine result lists
      testlist<-cbind(list,list2,list3)
      list_fin_mult[[k]]<-testlist
      save(list_fin_mult, file="...")
      }
```

Bibliography

Akaike, H. (1981). Likelihood of a model and information criteria. *Journal of Econometrics 16*(1), 3–14.

Anagnostopoulos, C., D. J. Hand, and N. M. Adams (2012). Measuring classification performance: The hmeasure package. Available online at `http://cran.r-project.org/web/packages/hmeasure/vignettes/hmeasure.pdf`.

Anderson, R. (2007). *The credit scoring toolkit: Theory and practice for retail credit risk management and decision automation.* Oxford: Oxford University Press.

Baesens, B., T. V. Gestel, S. Viaene, M. Stepanova, J. Suykens, and J. Vanthienen (2003). Benchmarking state-of-the-art classification algorithms for credit scoring. *Journal of the Operational Research Society 54*(6), 627–635.

Banasik, J. and J. Crook (2005). Credit scoring, augmentation and lean models. *Journal of the Operational Research Society 56*(9), 1072–1081.

Banasik, J., J. N. Crook, and L. C. Thomas (1999). Not if but when will borrowers default. *The Journal of the Operational Research Society 50*(12), 1185–1190.

Bartlett, P. L. and M. Traskin (2007). AdaBoost is consistent. *Journal of Machine Learning Research 8*(10), 2347–2368.

Basel Committee on Banking Supervision (2004). International convergence of capital measurement and capital standards: A revised framework. Available online at `http://www.bis.org/publ/bcbs107.pdf`.

Bradley, A. P. (1997). The use of the area under the ROC curve in the evaluation of machine learning algorithms. *Pattern Recognition 30*(7), 1145–1159.

Breiman, L. (1996). Bagging predictors. *Machine Learning 24*(2), 123–140.

Breiman, L. (1998). Arcing classifiers. *The Annals of Statistics 26*(3), 801–824.

Breiman, L. (1999). Prediction games and arcing algorithms. *Neural Computation 11*(7), 1493–1517.

Breiman, L. (2001a). Random forests. *Machine Learning 45*(1), 5–32.

Breiman, L. (2001b). Statistical modeling: The two cultures. *Statistical Science 16*(3), 199–215.

Breiman, L. (2002). Manual on setting up, using, and understanding random forests v3.1. Available online at `http://oz.berkeley.edu/users/breiman/Using_random_forests_V3.1.pdf`.

Breiman, L., J. H. Friedman, R. A. Olshen, and C. J. Stone (1984). *Classification and regression trees.* Belmont, California: Wadsworth International Group.

Brier, G. W. (1950). Verification of forecasts expressed in terms of probability. *Monthly Weather Review 78*(1), 1–3.

Brown, I. and C. Mues (2012). An experimental comparison of classification algorithms for imbalanced credit scoring data sets. *Expert Systems with Applications 39*(3), 3446–3453.

Bühlmann, P. (2006). Boosting for high-dimensional linear models. *The Annals of Statistics 34*(2), 559–583.

Bühlmann, P. and T. Hothorn (2007). Boosting algorithms: Regularization, prediction and model fitting. *Statistical Science 22*(4), 477–505.

Bühlmann, P. and B. Yu (2003). Boosting with the L2 loss: Regression and classification. *Journal of the American Statistical Association 98*(462), 324–339.

Calders, T. and S. Jaroszewicz (2007). Efficient AUC optimization for classification. In J. Kok, J. Koronacki, R. d. Lopez Mantaras, S. Matwin, D. Mladenič, and A. Skowron (Eds.), *Knowledge Discovery in Databases: PKDD 2007*, Volume 4702, pp. 42–53. Berlin and Heidelberg: Springer.

Cortes, C. and M. Mohri (2003). AUC optimization vs. error rate minimization. In S. Thrun, L. Saul, and B. Schölkopf (Eds.), *Advances in Neural Information Processing Systems 16: Proceedings of the 2003 Conference.* Cambridge: MIT Press.

Crook, J. and J. Banasik (2004). Does reject inference really improve the performance of application scoring models? *Journal of Banking & Finance 28*(4), 857–874.

Crook, J. N., D. B. Edelman, and L. C. Thomas (2007). Recent developments in consumer credit risk assessment. *European Journal of Operational Research 183*(3), 1447–1465.

Culp, M., K. Johnson, and G. Michailidis (2006). ada: An R package for stochastic boosting. *Journal of Statistical Software 17*(2), 1–27.

Culp, M., K. Johnson, and G. Michailidis (2007). ada. An R package for stochastic boosting: R package version 2.0-3. Available online at `http://CRAN.R-project.org/package=ada`.

DeLong, E. R., D. M. DeLong, and D. L. Clarke-Pearson (1988). Comparing the areas under two or more correlated receiver operating characteristic curves: A nonparametric approach. *Biometrics 44*(3), 837–845.

Dobra, A. and J. Gehrke (2001). Bias correction in classification tree construction. In *Proceedings of the Eighteenth International Conference on Machine Learning*, pp. 90–97. Morgan Kaufmann Publishers Inc.

Eguchi, S. and J. Copas (2002). A class of logistic–type discriminant functions. *Biometrika 89*(1), 1–22.

Fahrmeir, L., T. Kneib, and S. Lang (2009). *Regression: Modelle, Methoden und Anwendungen.* Heidelberg: Springer.

Fahrmeir, L., T. Kneib, S. Lang, and B. Marx (2013). *Regression: Models, Methods and Applications.* Springer.

Fawcett, T. (2006). An introduction to ROC analysis: ROC analysis in pattern recognition. *Pattern Recognition Letters 27*(8), 861–874.

Ferri, C., P. Flach, and J. Hernández-Orallo (2003). Improving the AUC of probabilistic estimation trees. In N. Lavrač, D. Gamberger, H. Blockeel, and L. Todorovski (Eds.), *Machine Learning: ECML 2003*, Volume 2837, pp. 121–132. Berlin and Heidelberg: Springer.

Ferri, C., P. Flach, J. Hernández-Orallo, and A. Senad (2005). Modifying ROC curves to incorporate predicted probabilities. In C. Ferri, N. Lachiche, S. Macskassy, and A. Rakotomamonjy (Eds.), *Proceedings of the 2nd workshop on ROC analysis in machine learning.* Available online at `http://users.dsic.upv.es/~flip/ROCML2005/papers/ferriCRC.pdf`.

Finlay, S. (2011). Multiple classifier architectures and their application to credit risk assessment. *European Journal of Operational Research 210*(2), 368–378.

Flach, P., J. Hernández-Orallo, and C. Ferri (2011). A coherent interpretation of AUC as a measure of aggregated classification performance. In *The 28th International Conference on Machine Learning.* Available online at `http://machinelearning.wustl.edu/mlpapers/paper_files/ICML2011Flach_385.pdf`.

Freund, Y., R. Iyer, R. E. Schapire, Y. Singer, and T. G. Dietterich (2004). An efficient boosting algorithm for combining preferences. *Journal of Machine Learning Research 4*(6), 933–969.

Freund, Y. and R. E. Schapire (1996). Experiments with a new boosting algorithm. In *Proceedings of the Thirteenth International Conference on Machine Learning.* San Francisco: Morgan Kaufmann.

Friedman, J., T. Hastie, and R. Tibshirani (2000). Additive logistic regression: A statistical view of boosting. *The Annals of Statistics 28*(2), 337–374.

Friedman, J. H. (1991). Multivariate adaptive regression splines. *The Annals of Statistics 19*(1), 1–67.

Friedman, J. H. (2001). Greedy function approximation: A gradient boosting machine. *The Annals of Statistics 29*(5), 1189–1232.

Frunza, M.-C. (2013). Computing a standard error for the gini coefficient: An application to credit risk model validation. *Journal of Risk Model Validation 7*(1), 61–82.

Geiger, C. and C. Kanzow (1999). *Numerische Verfahren zur Lösung unrestringierter Optimierungsaufgaben*. Berlin: Springer.

Han, A. K. (1987). Non-parametric analysis of a generalized regression model: The maximum rank correlation estimator. *Journal of Econometrics 35*(2–3), 303–316.

Hand, D. J. (2001). Modelling consumer credit risk. *IMA Journal of Management Mathematics 12*(2), 139–155.

Hand, D. J. (2009). Measuring classifier performance: A coherent alternative to the area under the ROC curve. *Machine Learning 77*(1), 103–123.

Hand, D. J. and N. M. Adams (2000). Defining attributes for scorecard construction in credit scoring. *Journal of Applied Statistics 27*(5), 527–540.

Hand, D. J. and W. E. Henley (1993). Can reject inference ever work? *IMA Journal of Management Mathematics 5*(1), 45–55.

Hand, D. J. and W. E. Henley (1997). Statistical classification methods in consumer credit scoring: A review. *Journal of the Royal Statistical Society. Series A 160*(3), 523–541.

Hand, D. J. and M. G. Kelly (2002). Superscorecards. *IMA Journal of Management Mathematics 13*(4), 273–281.

Hand, D. J. and R. J. Till (2001). A simple generalisation of the area under the ROC curve for multiple class classification problems. *Machine Learning 45*(2), 171–186.

Hanley, J. A. and B. J. McNeil (1982). The meaning and use of the area under a receiver operating characteristic (ROC) curve. *Radiology 143*(1), 29–36.

Hansen, B. E. (1997). Approximate asymptotic p values for structural-change tests. *Journal of Business & Economic Statistics 15*(1), 60–67.

Hastie, T. (2011). Generalized additive models: R package version 1.06.2. Available online at `http://CRAN.R-project.org/package=gam`.

Hastie, T., R. Tibshirani, and J. H. Friedman (2009). *The elements of statistical learning: Data mining, inference, and prediction* (2nd ed.). New York: Springer.

Hastie, T. J. and R. J. Tibshirani (1990). *Generalized additive models.* London: Chapman & Hall/CRC.

Henking, A., C. Bluhm, and L. Fahrmeir (2006). *Kreditrisikomessung: Statistische Grundlagen, Methoden und Modellierung.* Berlin: Springer.

Herschtal, A. and B. Raskutti (2004). Optimising area under the ROC curve using gradient descent. In *Proceedings of the twenty-first international conference on Machine learning*, New York, USA, pp. 49–56. ACM.

Hjort, N. L. and A. Koning (2002). Tests for constancy of model parameters over time. *Journal of Nonparametric Statistics 14*(1-2), 113–132.

Hofner, B., A. Mayr, N. Robinzonov, and M. Schmid (2014). Model-based boosting in R: a hands-on tutorial using the R package mboost. *Computational Statistics 29*(1-2), 3–35.

Hosmer, D. W. and S. Lemeshow (2000). *Applied logistic regression* (2nd ed.). New York: Wiley.

Hothorn, T., P. Bühlmann, T. Kneib, M. Schmid, and B. Hofner (2012). mboost. Model-based boosting: R package version 2.1-3. Available online at `http://CRAN.R-project.org/package=mboost`.

Hothorn, T., K. Hornik, C. Strobl, and A. Zeileis (2013). party. A laboratory for recursive partytioning: R package version 1.0-7. Available online at `http://CRAN.R-project.org/package=party`.

Hothorn, T., K. Hornik, and A. Zeileis (2006). Unbiased recursive partitioning: A conditional inference framework. *Journal of Computational & Graphical Statistics 15*(3), 651–674.

Janitza, S., C. Strobl, and A.-L. Boulesteix (2013). An AUC-based permutation variable importance measure for random forests. *BMC Bioinformatics 14*(119).

Jiang, W. (2004). Process consistency for adaboost. *The Annals of Statistics 32*(1), 13–29.

Komori, O. (2011). A boosting method for maximization of the area under the ROC curve. *Annals of the Institute of Statistical Mathematics 63*(5), 961–979.

Komori, O. and S. Eguchi (2010). A boosting method for maximizing the partial area under the ROC curve. *BMC Bioinformatics 11*(314).

Kraus, A. and H. Küchenhoff (2014). Credit scoring optimization using the area under the curve. *Journal of Risk Model Validation 8*(1), 31–67.

Kruppa, J., A. Schwarz, G. Arminger, and A. Ziegler (2013). Consumer credit risk: Individual probability estimates using machine learning. *Expert Systems with Applications 40*(13), 5125–5131.

Kullback, S. and R. A. Leibler (1951). On information and sufficiency. *The Annals of Mathematical Statistics 22*(1), 79–86.

Lagarias, J. C., J. A. Reeds, M. H. Wright, and P. E. Wright (1998). Convergence properties of the Nelder-Mead simplex method in low dimensions. *SIAM Journal of Optimization 9*, 112–147.

Lee, T.-S., C.-C. Chiu, Y.-C. Chou, and C.-J. Lu (2006). Mining the customer credit using classification and regression tree and multivariate adaptive regression splines. *Computational Statistics & Data Analysis 50*(4), 1113–1130.

Lessmann, S., H.-V. Seow, B. Baesens, and L. C. Thomas (2013). Benchmarking state-of-the-art classification algorithms for credit scoring: A ten-year update. In J. Crook (Ed.), *Proceedings of the Credit Scoring and Credit Control XIII.* Available online at http://www.business-school.ed.ac.uk/waf/crc_archive/2013/42.pdf.

Li, K.-C. and N. Duan (1989). Regression analysis under link violation. *The Annals of Statistics 17*(3), 1009–1052.

Liaw, A. and M. Wiener (2002). Classification and regression by random forest. *R News 2*(3), 18–22.

Liaw, A. and M. Wiener (2009). randomForest. Breiman and Cutler's random forests for classification and regression: R package version 4.5-34. Available online at http://CRAN.R-project.org/package=randomForest.

Lin, Y. and Y. Jeon (2006). Random forests and adaptive nearest neighbors. *Journal of the American Statistical Association 101*(474), 578–590.

Long, P. M. and R. A. Servedio (2007). Boosting the area under the ROC curve. In *21st Annual Conference on Neural Information Processing Systems.* Available online at http://www.cs.columbia.edu/~rocco/Public/nips07rocboost.pdf.

Ma, S. and J. Huang (2005). Regularized ROC method for disease classification and biomarker selection with microarray data. *Bioinformatics 21*(24), 4356–4362.

Mann, H. B. and D. R. Whitney (1947). On a test of whether one of two random variables is stochastically larger than the other. *The Annals of Mathematical Statistics 18*(1), 50–60.

Mayr, A., B. Hofner, and M. Schmid (2012). The importance of knowing when to stop. a sequential stopping rule for component-wise gradient boosting. *Methods Of Information In Medicine 51*(2), 178–186.

Miura, K., S. Yamashita, and S. Eguchi (2010). Area under the curve maximization method in credit scoring. *Journal of Risk Model Validation 4*(2), 3–25.

Nash, J. C. and R. Varadhan (2012). optimx. A replacement and extension of the optim() function: R package version 2011-8.2. Available online at `http://CRAN.R-project.org/package=optimx`.

Nelder, J. A. and R. Mead (1965). A simplex method for function minimization. *The Computer Journal 7*(4), 308–313.

Neyman, J. and E. S. Pearson (1933). On the problems of the most efficient tests of statistical hypotheses. *Philosophical Transactions of the Royal Society of London 231*, 289–338.

Pepe, M. S. (2003). *The statistical evaluation of medical tests for classification and prediction.* Oxford: Oxford University Press.

Pepe, M. S., T. Cai, and G. Longton (2006). Combining predictors for classification using the area under the receiver operating characteristic curve. *Biometrics 62*(1), 221–229.

Provost, F. and P. Domingos (2003). Tree induction for probability-based ranking. *Machine Learning 52*(3), 199–215.

Provost, F. and T. Fawcett (1997). Analysis and visualization of classifier performance: Comparison under imprecise class and cost distributions. In *Proceedings of the Third International Conference on Knowledge Discovery and Data Mining*, pp. 43–48. AAAI Press.

Quinlan, J. R. (1986). Induction of decision trees. *Machine Learning 1*(1), 81–106.

Quinlan, J. R. (1993). *C4.5 - programs for machine learning.* The Morgan Kaufmann series in machine learning. San Mateo and California: Kaufmann.

R Development Core Team (2012). *R: A Language and Environment for Statistical Computing.* Vienna and Austria: R Foundation for Statistical Computing. Available online at `http://www.R-project.org`.

Robinzonov, N. (2013). *Advances in boosting of temporal and spatial models.* Thesis. Ludwig-Maximilians-Universität München. Available online at `http://edoc.ub.uni-muenchen.de/15338/1/Robinzonov_Nikolay.pdf`.

Scheipl, F., L. Fahrmeir, and T. Kneib (2012). Spike-and-slab priors for function selection in structured additive regression models. *Journal of the American Statistical Association 107*(500), 1518–1532.

Schmid, M. and T. Hothorn (2008). Boosting additive models using component-wise P-splines. *Computational Statistics & Data Analysis 53*(2), 298–311.

Schmid, M., T. Hothorn, F. Krause, and C. Rabe (2012). A PAUC-based estimation technique for disease classification and biomarker selection. *Statistical Applications In Genetics And Molecular Biology 11*(5).

Sheng, V. S. and R. Tada (2011). Boosting inspired process for improving AUC. In P. Perner (Ed.), *Machine Learning and Data Mining in Pattern Recognition*, Volume 6871, pp. 199–209. Berlin and Heidelberg: Springer.

Sherman, R. P. (1993). The limiting distribution of the maximum rank correlation estimator. *Econometrica 61*(1), 123–137.

Stepanova, M. and L. Thomas (2002). Survival analysis methods for personal loan data. *Operations Research 50*(2), 277–289.

Strasser, H. and C. Weber (1999). On the asymptotic theory of permutation statistics. *Mathematical Methods of Statistics 2.*

Strobl, C., A.-L. Boulesteix, and T. Augustin (2007). Unbiased split selection for classification trees based on the gini index. *Computational Statistics & Data Analysis 52*(1), 483–501.

Strobl, C., A.-L. Boulesteix, T. Kneib, T. Augustin, and A. Zeileis (2008). Conditional variable importance for random forests. *BMC Bioinformatics 9*(307).

Strobl, C., A. L. Boulesteix, A. Zeileis, and T. Hothorn (2007). Bias in random forest variable importance measures: Illustrations, sources and a solution. *BMC Bioinformatics 8*(25).

Strobl, C., T. Hothorn, and A. Zeileis (2009). Party on! *The R Journal* (1/2), 14–17.

Strobl, C., J. Malley, and G. Tutz (2009). An introduction to recursive partitioning: Rationale, application, and characteristics of classification and regression trees, bagging, and random forests. *Psychological Methods 14*(4), 323–348.

Strobl, C. and A. Zeileis (2008). Danger: High power! Exploring the statistical properties of a test for random forest variable importance. *Proceedings of the 18th International Conference on Computational Statistics, Porto, Portugal.*

Swets, J. A. (1996). *Signal detection theory and ROC analysis in psychology and diagnostics: Collected papers.* Mahwah and New Jersey: Erlbaum.

Swets, J. A., R. M. Dawes, and J. Monahan (2000). Better decisions through science. *Scientific American 283*(4), 82–87.

Therneau, T. M. and B. Atkinson (1997). An introduction to recursive partitioning using the rpart routines. *Technical Report 61, Section of Biostatistics.* Available online at `http://www.mayo.edu/hsr/techrpt/61.pdf`.

Therneau, T. M. and B. Atkinson (2010). rpart. Recursive partitioning and regression trees: R package version 3.1-46. Available online at http://CRAN.R-project.org/package=rpart.

Thomas, L. C. (2000). A survey of credit and behavioural scoring: Forecasting financial risk of lending to consumers. *International Journal of Forecasting 16*(2), 149–172.

Thomas, L. C. (2009). *Consumer Credit Models: Pricing, Profit and Portfolios.* Oxford: Oxford University Press.

Thomas, L. C., R. W. Oliver, and D. J. Hand (2005). A survey of the issues in consumer credit modelling research. *Journal of the Operational Research Society 56*(9), 1006–1015.

Wang, G., J. Hao, J. Ma, and H. Jiang (2011). A comparative assessment of ensemble learning for credit scoring. *Expert Systems with Applications 38*(1), 223–230.

Wang, Z. (2011). Hingeboost: ROC-based boost for classification and variable selection. *The International Journal of Biostatistics 7*(13), 1–30.

Wilcoxon, F. (1945). Individual comparisons by ranking methods. *Biometrics Bulletin 1*(6), 80–83.

Wood, S. (2013). mgcv. Mixed GAM computation vehicle with GCV/AIC/REML smoothness estimation: R package version 1.7-22. Available online at http://CRAN.R-project.org/package=mgcv.

Wood, S. N. (2006). *Generalized additive models: An introduction with R.* Boca Raton, FL: Chapman & Hall/CRC.

Xiao, W., Q. Zhao, and Q. Fei (2006). A comparative study of data mining methods in consumer loans credit scoring management. *Journal of Systems Science and Systems Engineering 15*(4), 419–435.

Yan, L., R. Dodier, M. C. Mozer, and R. Wolniewicz (2003). Optimizing classifier performance via the Wilcoxon-Mann-Whitney statistic. In T. Fawcett and N. Mishra (Eds.), *Proceedings of the 20th international conference on machine learning,* Menlo Park, pp. 848–855. AAAI Press.

Zeileis, A. and K. Hornik (2007). Generalized M-fluctuation tests for parameter instability. *Statistica Neerlandica 61*(4), 488–508.

Zeileis, A., T. Hothorn, and K. Hornik (2008). Model-based recursive partitioning. *Journal of Computational & Graphical Statistics 17*(2), 492–514.

Zhang, H. and J. Su (2006). Learning probabilistic decision trees for AUC. *ROC Analysis in Pattern Recognition 27*(8), 892–899.

Eidesstattliche Versicherung

Hiermit erkläre ich an Eides statt, dass die Dissertation von mir selbstständig, ohne unerlaubte Beihilfe angefertigt ist.

München, den 10.03.2014

Anne Kraus